D1663982

Reviews of Physiology, Biochemistry and Pharmacology 149

Springer

Berlin
Heidelberg
New York
Hong Kong
London
Milan
Paris
Tokyo

Reviews of
149 Physiology Biochemistry and Pharmacology

Special Issue on Cytokines

Editors
S.G. Amara, Portland • E. Bamberg, Frankfurt
M.P. Blaustein, Baltimore • H. Grunicke, Innsbruck
R. Jahn, Göttingen • W.J. Lederer, Baltimore
A. Miyajima, Tokyo • H. Murer, Zürich
S. Offermanns, Heidelberg • N. Pfanner, Freiburg
G. Schultz, Berlin • M. Schweiger, Berlin

With 19 Figures and 3 Tables

 Springer

ISSN 0303-4240
ISBN 3-540-20213-7 Springer-Verlag Berlin Heidelberg New York

Library of Congress-Catalog-Card Number 74-3674

This work is subject to copyright. All rights are reserved, whether the whole or part of the material is concerned, specifically the rights of translation, reprinting, reuse of illustrations, recitation, broadcasting, reproduction on microfilms or in any other way, and storage in data banks. Duplication of this publication or parts thereof is permitted only under the provisions of the German Copyright Law of September 9, 1965, in its current version, and permission for use must always be obtained from Springer-Verlag. Violations are liable for prosecution under the German Copyright Law.

Springer-Verlag is a part of Springer Science+Business Media

springeronline.com

© Springer-Verlag Berlin Heidelberg 2004
Printed in Germany

The use of general descriptive names, registered names, trademarks, etc. in this publication does not imply, even in the absence of a specific statement, that such names are exempt from the relevant protective laws and regulations and therefore free for general use.
Product liability: The publishers cannot guarantee the accuracy of any information about dosage and application contained in this book. In every individual case the user must check such information by consulting the relevant literature.

Printed on acid-free paper – 14/3150 ag   5 4 3 2 1 0

D. Kamimura · K. Ishihara · T. Hirano

IL-6 signal transduction and its physiological roles: the signal orchestration model

Published online: 5 April 2003
© Springer-Verlag 2003

Abstract Interleukin (IL)-6 is a pleiotropic cytokine that not only affects the immune system, but also acts in other biological systems and many physiological events in various organs. In a target cell, IL-6 can simultaneously generate functionally distinct or sometimes contradictory signals through its receptor complex, IL-6Rα and gp130. One good illustration is derived from the in vitro observations that IL-6 promotes the growth arrest and differentiation of M1 cells through gp130-mediated STAT3 activation, whereas the Y759/SHP-2-mediated cascade by gp130 stimulation has growth-enhancing effects. The final physiological output can be thought of as a consequence of the orchestration of the diverse signaling pathways generated by a given ligand. This concept, *the signal orchestration model*, may explain how IL-6 can elicit proinflammatory or anti-inflammatory effects, depending on the in vivo environmental circumstances. Elucidation of the molecular mechanisms underlying this issue is a challenging subject for future research. Intriguingly, recent in vivo studies indicated that the SHP-2-binding site- and YXXQ-mediated pathways through gp130 are not mutually exclusive but affect each other: a mutation at the SHP-2-binding site prolongs STAT3 activation, and a loss of STAT activation by gp130 truncation leads to sustained SHP-2/ERK MAPK phosphorylation. Although IL-6/gp130 signaling is a promising target for drug discovery for many human diseases, the interdependence of each signaling pathway may be an obstacle to the development of a nonpeptide orally active small molecule to inhibit one of these IL-6 signaling cascades, because it would disturb the signal orchestration. In mice, a consequence of the imbalanced signals causes unexpected results such as gastrointestinal disorders, autoimmune diseases, and/or chronic

---

D. Kamimura · K. Ishihara · T. Hirano (✉)
Department of Molecular Oncology, Graduate School of Medicine,
Osaka University, Suita, Osaka, Japan
e-mail: hirano@molonc.med.osaka-u.ac.jp · Fax: +81-6-68793889

K. Ishihara · T. Hirano
Laboratory of Developmental Immunology, Graduate School of Frontier Biosciences,
Osaka University, Osaka, Japan

T. Hirano
Laboratory for Cytokine Signaling,
RIKEN Research Center for Allergy and Immunology, Yokohama, Japan

inflammatory proliferative diseases. However, lessons learned from IL-6 KO mice indicate that IL-6 is not essential for vital biological processes, but a significant impact on disease progression in many experimental models for human disorders. Thus, IL-6/gp130 signaling will become a more attractive therapeutic target for human inflammatory diseases when a better understanding of IL-6 signaling, including the identification of *the conductor* for gp130 signal transduction, is achieved.

Introduction

Interleukin (IL)-6 is a typical pleiotropic cytokine that modulates a variety of physiological events such as cell proliferation, differentiation, survival, and apoptosis. IL-6 plays roles in the immune system, the hematopoietic system, and inflammation. Furthermore, IL-6 has effects on the nervous system, endocrine system, bone metabolism, and other tissues and organ systems (Fig. 1). This review mainly focuses on the effects of IL-6 on the immune system and recent advances in understanding its intracellular signal transduction, but it also discusses other physiological roles of IL-6.

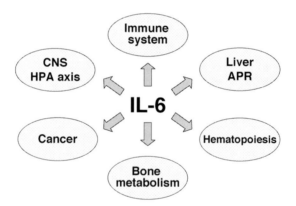

```
MNSFSTSAFG  PVAFSLGLLL  VLPAAFPAPV  PPGEDSKDVA  APHRQPLTSS   50
ERIDKQIRYI  LDGISALRKE  TCNKSNMCES  SKEALAENNL  NLPKMAEKDG  100
CFQSGFNEET  CLVKIITGLL  EFEVYLEYLQ  NRFESSEEQA  RAVQMSTKVL  150
IQFLQKKAKN  LDAITTPDPT  TNASLLTKLQ  AQNQWLQDMT  THLILRSFKE  200
FLQSSLRALR  QM                                              212
```

Fig. 1 The pleiotropy of IL-6 and its amino acid sequence. *Top*: IL-6 can modulate various biological events such as differentiation, proliferation, survival, and apoptosis in several organs and biological responses. APR, acute-phase response; CNS, central nervous system; HPA, hypothalamic-pituitary-adrenal axis. *Bottom*: the amino acid sequence of human IL-6. Amino acids are denoted with the single-letter code. The signal peptide is *shaded*, the four-cysteine motif is *boxed*, and the potential N-linked glycosylation sites are indicated with *asterisks*

Table 1 Summary of IL-6/IL-6Rα/gp130 homologs

Molecule	Synonyms	Species	Amino acid length		Homology to human[a]	Chromosome localization
			Precursor	Mature		
IL-6	IFNβ2, BSF-2, TRF, BCDF, BCDFII, HSF, 26kDa-protein, HPGF, MGI-2, IL-HP1, TRF-like factor	Human	212	185	–	7p21
		Mouse	211	187	42%	5
		Rat	211	nr	41%	4
		Pig	212	184	62%	9
		Cow	208	nr	53%	4
		Sheep	208	180	53%	4q1.3-q1.4
		HHV8	204	182	25%	–
IL-6Rα	CD126, gp80	Human	468	449	–	1q21
		Mouse	460	441	54%	3
		Rat	462	443	53%	2
gp130	CD130, IL-6ST, IL-6Rβ	Human	918	896	–	5q11
		Mouse	917	895	77%	13
		Rat	918	nr	78%	2q14-q16

[a] Overall homology at the amino acid level to human is shown
IFNβ2 interferon β2, *BSF-2* B cell stimulatory factor-2, *TRF* T cell-replacing factor, *BCDF* B cell-differentiation factor, *HPGF* hybridoma/plasmacytoma growth factor, *HSF* hepatocyte-stimulating factor, *MGI-2* monocyte/granulocyte inducer type-2, *IL-HP1* interleukin hybridoma plasmacytoma 1, *IL-6ST* IL-6 signal transducer, *nr* not reported

Historical overview

IL-6 is multifunctional, and thus several groups have independently identified and reported "IL-6" as different factors (see "synonyms" in Table 1). In 1980, Weissenbach et al. reported an inducible protein of 23 kDa–26 kDa, named interferon (IFN) β2, whose expression was stimulated in human fibroblasts by poly I:C, (Weissenbach et al. 1980). Independently, Hirano and his colleagues showed that the culture supernatant of purified-protein-derivative-stimulated pleural effusion cells from patients with pulmonary tuberculosis contains a potent activity for inducing B-cell growth and differentiation (Hirano et al. 1981). From these supernatants, they partially purified a factor, named TRF-like factor or B-cell differentiation factor II (BCDF-II), with a molecular weight of 22 kDa and an isoelectric point of 5–6, that was capable of inducing immunoglobulin (Ig) production in B cells (Teranishi et al. 1982; Hirano et al. 1984a; Hirano et al. 1984b). Hirano and colleagues then purified this factor and renamed it B cell stimulatory factor (BSF)p-2 (Hirano et al. 1985). In the same year, a factor termed 26-kDa-protein was found to be produced in IL-1-stimulated fibroblasts (Content et al. 1985), and a growth factor for mouse hybridoma was identified in the supernatant of human endothelial cells (Astaldi et al. 1980) and human monocytes. There was a race to molecularly clone these factors, and finally the cDNA sequence of BSFp-2 (BSF-2) was disclosed in 1986 (Hirano et al. 1986). That same year, Zilberstein et al. and May et al. reported the gene and cDNA structures of human IFN-β2 (May et al. 1986; Zilberstein et al. 1986), and the molecular cloning of 26-kDa-protein was also reported (Haegeman et al. 1986). By cDNA sequence comparison, these three molecules turned out to be identical. Subsequently, a hybridoma growth factor and a plasmacytoma growth factor were purified, and partial N-terminal amino acid sequences of these molecules revealed that they were the same as the BSF2/IFN-β2/26-kDa-protein. Thereafter, the cDNA cloning of a molecule identified as plasmacytoma and hybridoma growth factor and termed interleukin-HP1 (Van Snick et al. 1988), and a factor named hepatocyte-stimulating factor (Gauldie et al. 1987), which regulates the synthesis of acute-phase proteins, were reported. In 1989, the multifunctional factor with the various names was given a common designation, "interleukin-6" (Hirano et al. 1989; Le and Vilcek 1989; Sehgal et al. 1989; Hirano and Kishimoto 1990; Van Snick 1990; Hirano 1998).

Molecular aspects of IL-6

This section briefly summarizes molecular and biochemical characteristics of IL-6. To our knowledge, *IL-6* cDNAs from 15 vertebrates are compiled in the NCBI database. They are from human (database accession: M54894), mouse (J03783), rat (M26744), pig (AF309651), cow (X57317), sheep (X62501), rabbit (AF169176), dog (AF275796), cat (L16914), horse (AF041975), rhesus monkey (L26028), chicken (AJ309540), woodchuck (AF012908), beluga whale (AF076643), and red-crowned mangabey (L26032). Whether the molecules from other species are functionally homologous to human IL-6 has not been proven for all of them. However, the findings so far suggest that IL-6 is a universal factor among vertebrates.

Chromosome, gene, and cDNA

The human *IL-6* gene is approximately 5 kb long, consists of five exons with four introns (Yasukawa et al. 1987), and is located on chromosome 7p21 (Table 1; Bowcock et al. 1988). For the transcriptional control of *IL-6*, several potential promoter elements are found in the 5'-flanking region of the human *IL-6* gene. These include the glucocorticoid-responsive element, activating protein-1 (AP-1) binding site, multiple response element, c-fos serum-responsive element homolog, c-fos retinoblastoma control element homolog, cyclic AMP-responsive element, nuclear factor for IL-6 expression (NF-IL6, also known as the CCAAT/enhancer binding protein β)-binding site, and NF-κB binding site. The products of tumor-suppressor genes, p53 and retinoblastoma protein, are reported to repress the *IL-6* promoter activity contained in the sequence from nucleotides −225 to +13 (Santhanam et al. 1991).

The deduced amino acid sequence of human IL-6 consists of 212 amino acids with a signal peptide of 27 amino acids and two potential N-linked glycosylation sites (Fig. 1; Hirano et al. 1986). IL-6 homologs have been reported in more than a dozen vertebrate species. Table 1 lists a few examples with their chromosomal localization and the amino acid similarity of these homologs to human IL-6. In addition to vertebrates, Kaposi's sarcoma-associated herpes-like virus, also known as human herpes virus 8, encodes a functional IL-6 homolog called vIL-6 (Nicholas et al. 1997). The mouse IL-6 protein is 42% homologous to the human form and contains several potential O-linked glycosylation sites instead of the N-linked glycosylation site (Van Snick et al. 1988).

Protein

The human IL-6 protein has a molecular weight ranging from 21 kDa to 28 kDa with an isoelectric point of 5.4 (Fuller et al. 1987; Noda et al. 1991), consistent with the characteristics of the partially purified IL-6 (BCDF-II/TRF-like factor) previously reported by Teranishi et al. (Teranishi et al. 1982). IL-6 can undergo posttranscriptional modifications such as glycosylation (May et al. 1988a) and serine phosphorylation (May et al. 1988b). Because recombinant IL-6 protein produced by prokaryotes appears to be functional, its glycosylation seems not to be necessary for its biological activity (Tonouchi et al. 1988).

The X-ray crystal structure of human IL-6 shows a four-helix bundle, consisting of two pairs of antiparallel α-helices with up-up-down-down orientation (Somers et al. 1997), whose folding is conserved among cytokine family members. Based on the length of the helices, IL-6 is grouped into the "long chain" cytokines, which include growth hormone (GH), erythropoietin, and granulocyte colony-stimulating factor (G-CSF; reviewed by Bravo and Heath 2000).

Receptor for IL-6

This section describes the molecular features of the components of the IL-6 receptor, and also summarizes the characteristics of the IL-6 family cytokines that share one of the IL-6 receptor subunits, glycoprotein (gp)130.

Discovery of the IL-6 receptor

The first subunit of the IL-6 receptor (IL-6Rα) was molecularly cloned by an expression cloning system using biotinylated recombinant IL-6 as a probe. In addition, the same report proposed the presence of low- and high-affinity receptors for IL-6 and revealed that IL-6Rα was involved in both types of IL-6 receptor (Yamasaki et al. 1988). Subsequently, Taga et al. reported that the short cytoplasmic domain (82 amino acids) of IL-6Rα was not necessary for the IL-6-induced growth arrest of the M1 myeloid leukemic cell line. In addition, they showed that IL-6 stimulation triggered the association of IL-6Rα with another nonligand binding protein with a molecular mass of 130 kDa, termed gp130 (Taga et al. 1989). These results suggested the existence of a second, signal-transducing receptor for IL-6. The molecular cloning of the second IL-6 receptor component was achieved by Hibi et al. in 1990. As expected, the cytoplasmic domain of gp130 was found to carry several potential motifs for signal transduction (Hibi et al. 1990). Taken together, these studies showed that, although IL-6 can bind to IL-6Rα, gp130 is required for formation of the high-affinity receptor and generation of signal transduction.

Molecular aspects of the IL-6 receptor

As illustrated in Fig. 2, both IL-6Rα (also known as gp80 or CD126) and gp130 (also referred to as IL-6 signal transducer, IL-6Rβ, or CD130) contain an Ig-like domain and tandem fibronectin (FN) type-III domains including a four-cysteine motif and a tryptophan-serine-any tryptophan-serine (WSXWS) motif in the extracellular region. The four-cysteine and WSXWS motifs are responsible for the ligand binding, and thus are called the cytokine-binding module (CBM). Besides the CBM, gp130 has three additional FN type-III domains in its extracellular region. Both IL-6Rα and gp130 are grouped into the type-I cytokine receptor family, which also includes the receptors for prolactin, GH, many interleukins, leptin, erythropoietin, thrombopoietin, leukemia-inhibitory factor (LIF), oncostatin M (OSM), ciliary neurotrophic factor (CNTF), G-CSF, and granulocyte/macro-

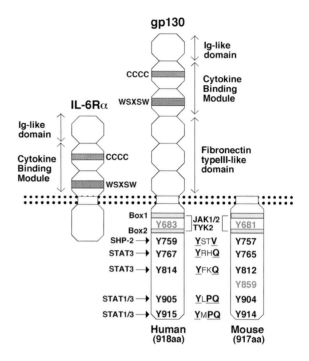

Fig. 2 IL-6Rα and gp130 structures. The structures for IL-6Rα and gp130 are shown. Amino acid residues are denoted with the *single-letter code*, followed by their position in the human and mouse gp130 amino acid sequences. JAK kinases are constitutively associated with gp130 through the Box domains. In response to IL-6 stimulation, SHP-2 and STAT molecules are recruited to respective tyrosine residues in gp130, indicated with *arrows* (see text for details). *CCCC* four-cysteine motif, *WSXWS* tryptophan-serine-any tryptophan-serine motif

phage-CSF (reviewed by Hirano et al. 1997; Hibi and Hirano 2001). Recently, a signal-transducing transmembrane receptor required for an invertebrate JAK/STAT pathway was identified in *Drosophila melanogaster*. The DOME/MOM protein has a similar structure to type-I cytokine receptors in vertebrates, in possessing a CBM containing a four-cysteine motif and a WSXWS-like sequence, and multiple FN type-III domains in its extracellular region, and a YXXQ motif (see below and the section entitled "Intracellular signal transduction pathways") in its cytoplasmic domain. Among vertebrate cytokine receptors, IL-6 family receptors, particularly LIFRβ and CNTFRα, are related to DOME/MOM (Brown et al. 2001; Chen et al. 2002). Thus, a signal-transducing mechanism is present in invertebrates that is similar to the IL-6-signaling system in vertebrates.

IL-6Rα is important for ligand binding, but it only has 82 amino acids in its cytoplasmic domain, indicating that it could play only a minor role, if any, in signal transduction (Taga et al. 1989). However, a recent study using a polarized epithelial cell line, Madin-Darby canine kidney cells, revealed that a membrane-proximal tyrosine-based YSLG motif and more distal dileucine-type LI motif in the cytoplasmic domain of IL-6Rα were involved in its asymmetrical expression on the basolateral side of the cells (Martens et al. 2000). In contrast, the cytoplasmic domain of gp130 contains several potential motifs for intracellular signaling, such as the YSTV sequence for SHP-2 (Src homology 2-containing tyrosine phosphatase 2) recruitment and YXXQ motifs (where X means any amino acids) for STAT (signal transducer and activator of transcription) activation (Fig. 2 and see the section entitled "Signal transduction" for details). Unlike many growth factor receptors, but common for cytokine receptors, gp130 does not have an intrinsic kinase domain (Hibi et al. 1990). Instead, like other cytokine receptors, the cytoplasmic domain of gp130 contains regions required for its association with a nonreceptor tyrosine kinase called Janus

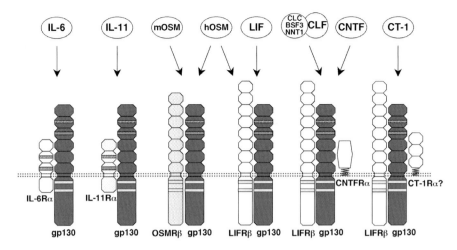

Fig. 3 gp130 is a shared signal transducer among IL-6 family cytokines. The receptor complex for IL-6 family cytokines consists of gp130 and a ligand-specific α chain, LIFRβ, or OSMRβ. gp130 is a common receptor component for all IL-6 family members. Human OSMβ (*hOSM*) uses both gp130/OSMRβ and gp130/LIFRβ while mouse OSM (*mOSM*) binds to gp130/OSMRβ only

kinase (JAK), by which downstream signaling cascades are initiated (see the section entitled "Signal transduction" for details).

The expression of gp130 is ubiquitous (Saito et al. 1992), while that of IL-6Rα is more restricted. IL-6Rα is found on hepatocytes, intestinal epithelial cells (Shirota et al. 1990), endocrine glands such as the pituitary and adrenal cortex (Bethin et al. 2000), and leukocytes, but not naïve B cells or certain cell lines. In addition, dexamethasone treatment up-regulates IL-6Rα expression in osteoblasts (Udagawa et al. 1995).

IL-6Rα has at least two types of soluble forms (sIL-6Rα) that are generated by proteolytic cleavage of the membrane-bound form or by alternative splicing of its mRNA. sIL-6Rα can act in an agonistic manner with IL-6 on cells expressing only gp130 (Receptor conversion model; reviewed by Hirano et al. 1997; Peters et al. 1998; Hirano and Fukada 2001; Jones et al. 2001). Thus, in association with sIL-6Rα, IL-6 can function in most parts of the body. gp130 also has a soluble form (sgp130) that is generated by alternative splicing of its mRNA. In contrast to sIL-6Rα, sgp130 acts in an antagonistic manner (Zhang et al. 1998; Tanaka et al. 2000; Jostock et al. 2001). Another splicing variant of gp130, termed gp130-RAPS, has been identified as an autoantigen in patients with rheumatoid arthritis. gp130-RAPS also antagonizes IL-6 activities in vitro (Tanaka et al. 2000).

IL-6 (gp130) family cytokines

Following the identification of gp130 as the IL-6 signal transducer, other cytokines were found to use gp130 as a receptor subunit as well. These include IL-11, CNTF, cardiotrophin-1 (CT-1), LIF, OSM, and a recently identified factor with three different names, cardiotrophin-like cytokine/novel neurotrophin-1/B cell-stimulating factor-3 (CLC/NNT1/BSF3). These cytokines are thus grouped into the IL-6- or gp130-family. The crystal and/or solution structures of IL-6, CNTF, OSM, and LIF have been resolved (Grotzinger et al. 1997; Deller et al. 2000).

Table 2 Phenotypes of IL-6 family cytokine mutants

Molecule	Phenotypes	References
IL-6 Tg	Plasmacytosis, mesangio-proliferative Glomerulonephritis (C57BL/6 background) Plasmacytoma (BALB/c background)	(Suematsu et al. 1989; Suematsu et al. 1992)
IL-6 Tg (CNS-specific)	Runting, tremor, ataxia, seizure neurodegeneration, astrocytosis, angiogenesis	(Campbell et al. 1993)
IL-6 Tg (Lung-specific)	Mononuclear cell infiltration in airways Airway eosinophilia ↓	(DiCosmo et al. 1994a; Wang et al. 2000a)
IL-6 Tg (Pancreas-specific)	Insulitis but not diabetes (NOD/F1) Delayed diabetes onset (NOD)	(DiCosmo et al. 1994b)
IL-6/sIL-6Rα Tg	Extramedullary hematopoiesis Tremor, gait abnormalities, paresis Hepatocyte hyperplasia, plasmacytoma	(Peters et al. 1997; Schirmacher et al. 1998; Brunello et al. 2000)
IL-6 KO	Resistance to bacterial and viral infection ↓ Thymocyte/peripheral T cell number ↓ APR↓, Mucosal IgA response ↓ Mature-onset obesity Resistant to bone loss by ovariectomy Leukocyte recruitment to inflammatory site ↓ Airway eosinophilia ↑ EAE ↓, CIA ↓, AIA ↓, EAMG↓	(Kopf et al. 1994; Poli et al. 1994; Ramsay et al. 1994; Romano et al. 1997; Ohshima et al. 1998; Wang et al. 2000a; Deng et al. 2002; Wallenius et al. 2002b)
IL-11 Tg (Lung-specific)	Airway eosinophilia ↓	(Wang et al. 2000b)
OSM Tg	Extrathymic T cell development Anti-dsDNA ↑, glomerulonephritis	(Clegg et al. 1999)
LIF Tg	Hypergammaglobulinemia, glomerulonephritis Disorganized thymic epithelium	(Shen et al. 1994)
LIF KO	Failure of blastocyst implantation Stem cell number ↓ HPA response ↓ Inflammatory cell infiltration↑, edema↑	(Stewart et al. 1992; Escary et al. 1993; Chesnokova et al. 1998; Zhu et al. 2001)
CT-1 KO	No obvious abnormalities up to 1 year of age Motoneuron cell death ↑ (E14-postnatal)	(Oppenheim et al. 2001)
CLF-1/NR6 KO	Perinatal death (within 24 h) Suckling defect	(Alexander et al. 1999)
CNTF KO	Born following the Mendelian rule Loss of motor neurons in adult mice EAE severity ↑	(Masu et al. 1993; Linker et al. 2002)
BSF3/NNT/CLC Tg	B cell hyperplasia Serum Ig level ↑, anti-dsDNA ↑ Immunotactoid glomerulopathy	(Senaldi et al. 2002)

Ag antigen, *CIA* collagen-induced arthritis, *CNS* central nervous system, *ds* double-strand, *EAE* experimental autoimmune encephalomyelitis, *EAMG* experimental autoimmune myasthenia gravis, *HPA* hypothalamus/pituitary/adrenal, *KO* knock-out, *Tg* transgenic

All IL-6 family cytokines share gp130 as one of their receptor components, as illustrated in Fig. 3. Like the IL-6 situation, IL-11 and CNTF bind to a ligand-specific receptor, IL-11Rα and CNTFRα, respectively. CNTFRα is unique in that it is a glycosylphosphatidylinositol (GPI)-anchored protein. There is evidence that CT-1 also has a ligand-specific receptor, CT-1Rα, which associates with gp130 to form a high-affinity receptor (Robledo et al. 1997). On the other hand, the recently added IL-6-family member, CLC/NNT1/

Table 3 Phenotypes of IL-6 family cytokine receptor mutants

Molecule	Phenotypes	References
gp130 KO	E12.5—perinatal death Hematopoiesis ↓ Osteoclast number ↑(ICR background)	(Yoshida et al. 1996; Kawasaki et al. 1997)
gp130 KO (IFN-inducible)	Resistance to bacterial and viral infection ↓ Thymocyte/peripheral T cell number ↓ Acute-phase response ↓ Thrombopoiesis ↓ Life span ↓ (emphysema)	(Betz et al. 1998)
gp130 KO (Heart-specific)	Aortic pressure overload-induced dilated cardiomyopathy, myocyte apoptosis ↑	(Hirota et al. 1999)
gp130^{F759} KI (SHP-2 signal defect)	TD response ↑ Splenomegaly, lymphadenopathy Th1 skewed (IFN-γ ↑, IL-4 ↓) Resistance to *Listeria* infection ↓ Rheumatoid arthritis-like joint disease	(Ohtani et al. 2000; Atsumi et al. 2002; Kamimura et al. 2002)
gp130^{FXXQ} KI (STAT3 signal defect)	Perinatal death TD response ↓(fetal liver chimera)	(Ohtani et al. 2000)
gp130^{D} KI (All signal defect)	Perinatal death TD response ↓ (fetal liver chimera)	(Ohtani et al. 2000)
gp130^{757F} KI (SHP-2 signal defect)	Gastric adenoma Completely resistant to DSS-induced colitis	(Tebbutt et al. 2002)
gp130$^{\Delta STAT}$ KI (truncation, STAT3 signal defect)	Body weight ↓, trunk length ↓, life span ↓ Gastrointestinal ulceration Degenerative joint disease Sensitivity to DSS-induced colitis↑	(Ernst et al. 2001; Tebbutt et al. 2002)
DN gp130 Tg	Thymocyte/lymphocyte number ↓ TD response ↓ Pressure overload-induced cardiac hypertrophy ↓	(Kumanogoh et al. 1997; Uozumi et al. 2001)
gp130$^{\Delta STAT}$/LIFR-KO (double heterozygotes)	EAE severity ↑	(Butzkueven et al. 2002)
IL-11Rα KO	Normal hematopoiesis Female infertility	(Nandurkar et al. 1997; Robb et al. 1998)
CNTFRα KO	Perinatal death, lack of feeding Severe motor neuron deficits	(DeChiara et al. 1995)
LIFRβ KO	Perinatal death Osteoclast number ↑ Astrocyte number ↓	(Ware et al. 1995)

DN dominant negative, *EAE* experimental autoimmune encephalomyelitis, *KI* knock-in, *KO* knock-out, *TD* thymus dependent, *Tg* transgenic

BSF3, forms a heterodimer with cytokine-like factor 1 (CLF-1, also known as NR6), which is a soluble protein that belongs to the type-I cytokine receptor family but lacks a transmembrane region (Elson et al. 1998; Alexander et al. 1999). Interestingly, the expression of CLF-1 is required for CLC/NNT1/BSF3 secretion, and the heterodimer acts on cells expressing functional CNTF receptors (Elson et al. 2000). Because phenotypes between CNTF knock-out (KO) and CNTFRα KO mice are incompatible with each other (see Tables 2 and 3), the existence of a second ligand for CNTFR has been proposed. In addition, CLF-1/NR6 KO mice show a suckling defect and die within 24 h after birth, similar to CNTFRα KO mice (Alexander et al. 1999). Thus, the complex of CLC/NNT1/BSF3 and CLF-1 is the best candidate for the second ligand of CNTFR (Elson et al. 2000; Lelievre et al. 2001). The rest of the IL-6 family members, LIF and OSM, bind to a receptor complex consisting of gp130 and another gp130-related signal transducing receptor, LIFRβ and OSMRβ, respectively. Human OSM can bind either gp130:LIFRβ or gp130:OSMRβ, whereas mouse OSM binds gp130:OSMRβ only (Ichihara et al. 1997).

With the shared use of the signal transducer gp130, the biological functions of IL-6 family cytokines are largely overlapping. For example, all known IL-6 family cytokines can induce the production of acute-phase proteins (APPs). IL-6, LIF, OSM, and CT-1 induce macrophage differentiation of the mouse leukemic cell line M1 in the absence of any additional factors. On the other hand, nonredundant activities of the IL-6 family cytokines have been disclosed by gene targeting studies. As summarized in Tables 2 and 3, female IL-11Rα KO mice are infertile due to defective decidualization, and blastocysts from LIF KO mice are unable to be implanted, demonstrating that IL-11 and LIF possess nonredundant effects on female reproduction. By contrast, IL-6 KO mice exhibit no overt developmental defects and are apparently healthy and fertile, demonstrating IL-6 is not essential to life; however, IL-6 has a great impact on immune systems, as discussed in the next section.

Physiological functions of IL-6

Many cell types are reported to produce IL-6; these include T cells, B cells, polymorphonuclear cells, eosinophils, monocyte/macrophages, mast cells, dendritic cells, chondrocytes, osteoblasts, endothelial cells, skeletal and smooth muscle cells, islet β cells, thyroid cells, fibroblasts, mesangial cells, keratinocytes, and certain tumor cells. In addition, adipose tissue is a source of IL-6. Microglial cells and astrocytes are also IL-6 producers. Taking the widespread distribution of gp130 and the shedding of sIL-6Rα into consideration, it is easy to imagine how IL-6 can function in a wide variety of systems in the body, as described in this section.

Immune system

Since IL-6 was originally identified as a factor promoting Ig secretion (then called BSF-2, the section entitled "Historical overview"), it has been known to act on B cells. IL-6 enhances the production of IgM, IgG, and IgA in B cells activated with *Staphylococcus aureus* Cowan I or pokeweed mitogen. Conversely, an anti-IL-6 antibody inhibits pokeweed mitogen-induced Ig production from peripheral blood mononuclear cells without affecting cell proliferation, indicating an absolute requirement for IL-6 for antibody production in B cells (Muraguchi et al. 1988). Mouse IL-6 also acts on B cells activated with anti-Ig or dextran sulfate (Vink et al. 1988). For the mechanism of IL-6 action on B cells, a study using primary B cells from a gp130 knock-in strain (see the sections entitled "Intracellular signal transduction pathways" and "The signal orchestration model" for details) demonstrated that the promoting effect of IL-6 on Ig production is dependent on gp130 YXXQ-mediated signaling, most likely through STAT3 activation (Ohtani et al. 2000). Many studies using IL-6-overexpressing transgenic (Tg) or IL-6 KO mice have revealed important in vivo roles of IL-6 in the immune system (Table 2). The constitutive overexpression of IL-6 in mice leads to the development of mesangio-proliferative glomerulonephritis with IgG1 plasmacytosis in the C57BL/6 background (Suematsu et al. 1989), and the development of transferable plasmacytoma with chromosomal translocation t(12;15) in the BALB/c background (Suematsu et al. 1992). Kopf et al. was the first to report IL-6 KO mice, in 1994. Although the serum Ig levels in IL-6 KO mice are indistinguishable from those in wild-type animals, the antivirus IgG (Kopf et al. 1994) and mucosal IgA antibody (Ramsay et al. 1994) responses are severely impaired in IL-6 KO mice. These results demonstrate that IL-6 acts on B cells to promote Ig production and is a growth factor for plasmacytoma in vivo.

In T cells, IL-6 confers supportive but significant effects on proliferation, survival, and type-1 helper T-cell (Th1)/Th2 responses. When T cells are stimulated with anti-CD3 and IL-6, their proliferation is significantly enhanced compared with anti-CD3 stimulation alone. In addition, IL-6 prevents anti-CD3-induced apoptosis in T cells. These effects are mediated by STAT3, because they are lost in T-cell-specific STAT3 KO mice (Takeda et al. 1998). The significance of IL-6-mediated STAT3 activation in blocking T-cell apoptosis was also shown in a recent study using a knock-in mouse strain defective in STAT3 autoregulation (Narimatsu et al. 2001). In fact, thymocyte and peripheral T-cell numbers are consistently reduced in IL-6 KO mice and in transgenic mice containing a dominant-negative form of gp130, compared with wild-type animals (Kopf et al. 1994; Kumanogoh et al. 1997). IL-6 also affects the Th1/Th2 balance. Rincon et al. provided evidence that IL-6 directs Th2 differentiation by two distinct mechanisms. When IL-6 is added to a culture under conditions that induce Th differentiation, T cells produce more Th2 cytokine (IL-4) and less Th1 cytokine (IFN-γ) than a culture lacking IL-6. The Th2-promoting effect of IL-6 is mediated by IL-4, because an anti-IL-4 antibody can neutralize this action (Rincon et al. 1997). Recently, the molecular mechanism underlying this phenomenon was resolved. Diehl et al. found that IL-6 upregulates the NFAT (nuclear factor of activated T cells) transcriptional activity by increasing the levels of NFATc2. In T cells from transgenic mice expressing a dominant-negative form of NFAT or in NFATc2 KO mice, the ability of IL-6 to promote Th2 differentiation is diminished. These results suggest that IL-6 enhances NFAT activity in naive CD4 T cells, leading to an upregulation of endogenous IL-4 production. IL-4 then provides signals for Th2 differentiation (Diehl et al. 2002). On the other hand, IL-6 is also able to interfere with IL-12-mediated Th1 differentiation by a mechanism distinct from that for the Th2 enhancement. In this case, SOCS1 (suppressor of cytokine signaling 1) is induced by IL-6 and inhibits IFN-γR signaling in T cells. The inhibition of IFN-γR-mediated signals by IL-6 prevents the autoregulation of the *IFN-γ* gene expression, thereby preventing Th1 differentiation. Thus, IL-6, probably produced by antigen-presenting cells, is a key modulator of Th1/Th2 differentiation (Diehl et al. 2000). In contrast to these observations, however, it is also reported that T cells or lymph node cells from IL-6 KO mice immunized with an antigen produce more IL-4 when re-stimulated in vitro with the same antigen, suggesting that the absence of IL-6 predisposes the Th balance to the Th2-type (Tanaka et al. 2001). Similarly, when experimental models for rheumatoid arthritis are induced in IL-6 KO mice, IL-6 KO T cells show cytokine profiles that are shifted to those of Th2 cells (i.e., increased IL-4 production; Ohshima et al. 1998; Sasai et al. 1999). On the other hand, another report shows that the absence of IL-6 does not affect Th2 differentiation in vivo (La Flamme and Pearce 1999). Furthermore, Ohtani et al. reported that T cells defective in gp130-mediated STAT3 activation produce a lower amount of IFN-γ than do wild-type T cells. In contrast, CD4-positive T cells in which the gp130-mediated SHP-2/ERK MAPK (extracellular signal-regulated kinase, mitogen-activated protein kinase) cascade is impaired, exhibit more IFN-γ and less IL-4 compared with the same cells from wild-type mice, suggesting that SHP-2 and STAT3 signals relayed through gp130 reciprocally regulate the balance of Th1/Th2 cytokine production. Taking these observations together, although it is clear that IL-6 has a modulatory effect on the Th1/Th2 balance or responses, its effect on the differentiation of Th cells is still controversial.

IL-6 also affects the differentiation of professional antigen-presenting cells such as macrophages and dendritic cells (DCs). When stimulated with GM-CSF and IL-4, human peripheral blood monocytes differentiate into DCs. The addition of IL-6 to the culture system switches the differentiation of monocytes from DCs to macrophages. This switch in differentiation results from the IL-6-induced upregulation of M-CSF receptors on monocytes (Chomarat et al. 2000; Mitani et al. 2000). In contrast, the stimulation of gp130 on DCs by an agonistic monoclonal antibody promotes the differentiation and maturation of these cells in response to GM-CSF plus IL-4, which is associated with the upregulation of chemokine production and costimulatory molecules (Wang et al. 2002). The idea that gp130 stimulatory cytokines other than IL-6 possess the differentiation-promoting effects is a likely interpretation of these contradictory results.

IL-6 also modulates leukocyte recruitment. In a well-known experimental system, the injection of carageenan into an artificially created subcutaneous dorsal air-pouch in mice induces local inflammation. In IL-6 KO mice subjected to this procedure, the inflammatory responses, including the numbers of infiltrating leukocytes and chemokine levels, are reduced. Because the in vitro chemotactic response of polymorphonuclear cells and macrophages from IL-6 KO mice is normal, the defective migratory responses in these mice is not an autonomous effect of the leukocytes but results from a reduced expression of chemokines and integrins on endothelial cells (Romano et al. 1997). The lung-specific or pancreatic islet β cell-specific overexpression of IL-6 in mice results in the infiltration of mononuclear cells into the affected areas (DiCosmo et al. 1994a; DiCosmo et al. 1994b). These results further support the functional effect of IL-6 on leukocyte recruitment.

Several reports indicate that IL-6 plays an important role in the host response to bacterial and viral infection. IL-6 KO mice cannot efficiently control vaccinia virus and *Listeria monocytogenesis* infections (Kopf et al. 1994). IL-6 KO mice are also highly susceptible to infection by *Escherichia coli* (Dalrymple et al. 1996) and *Candida albicans* (Romani et al. 1996). Conversely, the injection of recombinant IL-6 into mice rendered them more resistant to *Listeria* infection (Liu et al. 1995). A knock-in strain carrying a mutation at Y759 of gp130 ($gp130^{F759/F759}$, see the section entitled "Signal transduction" for a description of these mice) also shows enhanced susceptibility to *Listeria* infection, suggesting that gp130-mediated SHP-2/ERK MAPK signals are critical for bacterial resistance in vivo (Kamimura et al. 2002).

Acute-phase reaction

The acute-phase reaction (APR) is rapidly induced by inflammation associated with infection, injury, and other factors. This reaction serves to neutralize pathogens and prevent further invasion by them, and also to minimize tissue damage, thereby promoting the body's recovery from the unwanted inflammatory state. The APR consists of fever, an increase in vascular permeability, and the production of acute-phase proteins (APPs) by hepatocytes. The APPs are divided into two groups based on the cytokines that regulate them. IL-6 (IL-6 family cytokines) directly upregulates the mRNA expression of type-II APPs through STAT3 activation. IL-6 also contributes to the increase in type-I APP levels, which are mainly regulated by IL-1 (reviewed by Heinrich et al. 1990; Baumann and Gauldie 1994; Moshage 1997). In IL-6 KO mice, the production of APPs induced by turpentine, *Listeria*, or lipopolysaccharide injection is lower than that observed in wild-type mice (Kopf et al. 1994).

Nervous and endocrine systems

IL-6 mRNA can be detected by in situ hybridization in the hippocampus, hypothalamus, and subcortical structures of the rat brain. In addition, sIL-6Rα is detectable in human cerebrospinal fluid. These findings imply a functional role of IL-6 in the central nervous system (CNS). In vitro, IL-6 can induce neurite outgrowth in the rat pheochromocytoma cell line (PC12) when the cells are pretreated with nerve-growth factor. This effect is mediated by IL-6-induced, gp130 Y759-derived ERK MAPK activation (Ihara et al. 1997). The CNS-specific overexpression of IL-6 in mice results in the development of reactive gliosis, and these mice display significant neurodegeneration, which causes motor problems such as ataxia, seizures, and tremors (Campbell et al. 1993). In addition, IL-6 KO mice show a delayed recovery of sensory functions after crush lesions of the sciatic nerve (Zhong et al. 1999), demonstrating the functional effects of IL-6 on the CNS.

IL-6 and IL-6 family cytokines are reported to modulate the hypothalamic-pituitary-adrenal (HPA) axis. Injection of recombinant IL-6 stimulates the release of adrenocorticotrophic hormone in a manner independent of the action of corticotropin-releasing hormone. The expression of IL-6Rα on pituitary corticotrophs and in the adrenal cortex supports the direct action of IL-6 on hormone release. (reviewed by Bethin et al. 2000)

A recent report uncovered another interesting role of IL-6, in the modulation of body mass in adult animals. Wallenius et al. demonstrated that IL-6 KO mice develop mature-onset obesity: IL-6 KO mice gain 20% more body weight than wild-type animals at around 9 months of age, which is mainly due to an increase in subcutaneous fat. The obesity in IL-6 KO mice is accompanied by disturbed carbohydrate and lipid metabolism, an elevated leptin level, and reduced responsiveness to leptin treatment. In addition, intracerebroventricular, but not intraperitoneal, injection of IL-6 into these rats increases the energy expenditure, and eventually reduces body fat, indicating a centrally acting antiobesity effect of IL-6 (Wallenius et al. 2002a; Wallenius et al. 2002b).

Cancer

Many tumors, including Kaposi's sarcoma (Miles et al. 1990), melanoma (Molnar et al. 2000), multiple myeloma (Kawano et al. 1988; Frassanito et al. 2001; see also reviews in Hirano 1991; Klein et al. 1995; Hirano et al. 2000), and prostate cancer (Smith et al. 2001; Ueda et al. 2002) produce IL-6, which can act as an autocrine and/or paracrine growth factor for the neoplasm. In fact, when human multiple myeloma cells are injected into immunodeficient *scid* mice, treatment with a humanized anti-IL-6Rα monoclonal antibody suppresses the tumor-associated abnormalities and prolongs the lifespan of the tumor-bearing mice (Tsunenari et al. 1997). In addition, an anti-IL-6 neutralizing antibody, a receptor antagonistic IL-6 mutant called Sant7, or an antisense oligodeoxynucleotide against gp130 can all suppress the effect of IL-6 on the proliferation and the enhancement of drug resistance in the human prostate carcinoma PC-3 cell line (Borsellino et al. 1999).

Bone metabolism

IL-6 is implicated in osteoclastogenesis. Osteoclast formation is enhanced by ovariectomy in mice, and this effect is negated when an anti-IL-6 antibody or 17β-estradiol is administered to the mice (Jilka et al. 1992). IL-6 plus sIL-6Rα strikingly triggers osteoclast formation in a coculture system containing mouse bone marrow cells and osteoblastic cells. The ability of IL-6 to induce osteoclast differentiation is dependent on the signal transduction in the osteoblastic cells, but not in osteoclast progenitors (Udagawa et al. 1995). Consistent with these findings, IL-6 KO mice are resistant to the bone loss induced by ovariectomy (Poli et al. 1994). These findings suggest an important role for IL-6 in the osteoporosis found in postmenopausal women (Jilka et al. 1992) and patients with rheumatoid arthritis (Rifas 1999).

Hematopoiesis

Most primitive hematopoietic progenitors are positive for gp130 expression, and about 30%–50% of the population expresses IL-6Rα. Stimulation of gp130 with IL-6 and sIL-6Rα induces significant expansion of human CD34-positive cord blood cells in combination with stem cell factor (reviewed by Peters et al. 1998). Megakaryopoiesis is stimulated by IL-6 in concert with stem cell factor and thrombopoietin. In fact, IL-6 KO mice have a reduced number of megakaryocyte progenitors (Bernad et al. 1994). In addition, IL-6 is involved in granulopoiesis. IL-6/G-CSFR double KO mice display neutropenia that is more severe than that observed in G-CSFR KO mice. Moreover, the injection of IL-6 into G-CSFR KO mice improves the granulopoiesis (Liu et al. 1997). IL-6 can induce growth arrest and differentiation in human (U937) and mouse (M1) myeloid cell lines. The murine M1 cells differentiate into macrophages in response to IL-6; this is mediated by IL-6-induced STAT3 activation (Yamanaka et al. 1996), as described later in the section entitled "The signal orchestration model".

Mice overexpressing both IL-6 and IL-6Rα show massive extramedullary hematopoiesis in the spleen and liver (Schirmacher et al. 1998; Table 2), which supports the above observations.

Relevance of IL-6 to human diseases

The IL-6 level in the circulation stays low under normal conditions in healthy, young individuals. However, IL-6 production is rapidly induced in the course of acute inflammatory reactions associated with injury, trauma, stress, infection, and other situations. In addition, aging also influences the production of IL-6. With advancing age, plasma IL-6 levels increase, a change that is explained at least in part by age-associated diseases. Because estrogens and androgens are known to repress IL-6 expression, however, a decrease in the sex hormones with age may also contribute to the increase in IL-6 (Ershler and Keller

2000). As the source of IL-6, not only immune cells, but a variety of other cell types, such as muscle cells (Pedersen et al. 2001), adipocytes (Yudkin et al. 2000), hepatocytes, microglial cells, and astrocytes produce IL-6. The IL-6 receptor component, gp130, is broadly distributed in the body (see the section entitled Molecular aspects of the IL-6 receptor). Hence, a dysregulated, high-level production of IL-6 that has pleiotropic effects could induce an undesired inflammatory state in many organs, a condition that can cause various diseases. In fact, a number of reports implicate IL-6 in the pathogenesis of many human disorders, including Alzheimer's disease (O'Barr and Cooper 2000; Papassotiropoulos et al. 2001), bronchial asthma (Yokoyama et al. 1995; Wong et al. 2001), cardiac myxoma (Hirano et al. 1987), Castleman's disease (Yoshizaki et al. 1989), inflammatory bowel disease (Holtkamp et al. 1995), multiple myeloma (Kawano et al. 1988), multiple sclerosis (Stelmasiak et al. 2000), rheumatoid arthritis (Hirano et al. 1988), Sjögren's syndrome (Grisius et al. 1997), systemic lupus erythematosus, (Stuart et al. 1995), type-II diabetes mellitus (Pradhan et al. 2001), and others. Because a variety of useful animal models for immunological disorders are available, many investigators have attempted to clarify the role of IL-6 in the pathogenic mechanisms underlying disease development. Evidence from in vitro studies indicates that IL-6 has so-called assisting effects on the functions of a variety of immune cells. However, as listed below, IL-6 appears to be essential for the progression of experimentally induced-immunological disorders in animals, making IL-6 an attractive therapeutic target.

1. Inflammatory bowel diseases (IBD)
   IBD commonly encompasses two chronic, tissue-destructive clinical entities, Crohn's disease (CD) and ulcerative colitis (UC), which are possibly caused by an immunological hypersensitivity to commensal gut bacteria. Suzuki and colleagues found that among the STAT family members, STAT3 is most strongly tyrosine-phosphorylated in colon tissue extracts from patients with UC and CD, and from mice suffering from experimentally induced colitis. The development of the experimental colitis as well as the STAT3 activation in the colon is significantly reduced in IL-6 KO mice (Suzuki et al. 2001). In an independent study, Atreya et al. showed that in UC patients, lamina propria (LP) cells produce large amounts of IL-6 and sIL-6Rα, and that STAT3 activation is also observed in these cells. Using various animal models of UC, this group demonstrated that the neutralization of IL-6 signaling by an anti-IL-6Rα antibody or gp130-Fc fusion protein results in a suppression of colitis activity, which is correlated with an induction of apoptosis in LP T cells (Atreya et al. 2000). These findings suggest that IL-6 and sIL-6Rα-mediated STAT3 activation plays an important role in the perpetuation of colitis.
2. Multiple sclerosis
   Experimental autoimmune encephalomyelitis (EAE) is an animal model for a demyelinating disease, multiple sclerosis. EAE can be induced in mice by immunization with myelin components such as myelin oligodendrocyte glycoprotein (MOG). Several reports demonstrated that IL-6 KO mice are resistant to the MOG-induced EAE, compared with wild-type mice (Eugster et al. 1998; Okuda et al. 1998; Samoilova et al. 1998). The resistance to EAE of IL-6 KO mice is associated with a deficiency of MOG-specific T cells, which can differentiate into either Th1 or Th2 type effecter cells in vivo (Samoilova et al. 1998). Histologically, no infiltration of inflammatory cells is observed in the CNS of IL-6 KO mice. This is due to a decreased expression of endothelial adhesion molecules, VCAM-1 and ICAM-1, which are upregulated in the CNS of wild-type mice that show the symptoms of MOG-induced EAE (Eugster et al.

1998). Further analyses are required to understand the mechanisms underlying how IL-6 regulates the immune responses in EAE.

3. Myasthenia gravis

   Myasthenia gravis (MG) is an antibody-mediated autoimmune neuromuscular disease in which the acetylcholine receptor (AChR) in the neuromuscular junction is destroyed by anti-AChR and the complement system. An animal model of MG, called experimental autoimmune myasthenia gravis (EAMG), can be induced in vertebrates by immunization with *Torpedo californica* AChR in complete Freund's adjuvant (reviewed by Christadoss et al. 2000). When EAMG is induced in IL-6 KO mice, only 25% of the animals develop the clinical manifestations of EAMG, while 83% of wild-type controls develop the symptoms. The EAMG resistance in IL-6 KO mice is associated with a significant reduction in anti-AChR antibody levels, the AChR-specific proliferative response of T cells, and germinal center formation. This is the first genetic evidence that IL-6 is involved in the pathogenesis of MG (Deng et al. 2002).

4. Bronchial asthma

   Asthma is a chronic respiratory disorder characterized by reversible airflow obstruction and airway inflammation, persistent airway hyperreactivity, and airway remodeling (reviewed by Maddox and Schwartz 2002). In sharp contrast to observations from other disease models, the symptoms of antigen-induced pulmonary eosinophilia, an animal model for bronchial asthma, are exaggerated in IL-6 KO mice. Augmented eosinophilic infiltration in the bronchoalveolar lavage fluid, enhanced airway responsiveness to methacholine, and increased levels of chemokines and Th2-type cytokines are observed in IL-6 KO mice. On the other hand, these symptoms are reduced in mice overexpressing IL-6 under the lung-specific promoter, CC10 (Wang et al. 2000a). In addition, aerosol delivery of lipopolysaccharide into mice causes the production of tumor-necrosis factor α and a chemokine, MIP2, and neutrophil accumulation in the bronchoalveolar lavage fluid. This acute lung inflammation is enhanced in IL-6 KO mice (Xing et al. 1998). These results indicate that IL-6 can exhibit anti-inflammatory effects under certain conditions.

5. Rheumatoid arthritis

   Rheumatoid arthritis (RA) is a heterogeneous, chronic joint disease characterized by leukocyte invasion and synoviocyte activation, followed by cartilage and bone destruction. It has properties of both autoimmune and chronic proliferative inflammatory diseases (reviewed by Feldmann et al. 1996; Hirano 1998). The possible involvement of IL-6 in RA was first demonstrated by the high levels of IL-6 detected in synovial fluid from the joints of patients with active RA (Hirano et al. 1988). The involvement of IL-6 in the pathogenesis of RA was then revealed in studies using several animal models of this disease. One of them is type-II collagen-induced arthritis (CIA). When CIA is elicited in IL-6 KO mice backcrossed to a susceptible genetic background, these mice show a delay of onset and reduced severity of clinical symptoms of CIA (Alonzi et al. 1998; Sasai et al. 1999). The humoral and cellular responses to type-II collagen in the IL-6 KO mice are about half those seen in wild-type mice. In addition, the Th cellular responses in IL-6 KO mice are shifted to the Th2 type, as judged by the enhanced production of IL-4 and IL-10 in response to concanavalin A stimulation (Sasai et al. 1999). Antigen-induced arthritis (AIA) is another experimental model for RA. When AIA is induced, the articular cartilage is completely destroyed in genetically susceptible wild-type mice, whereas IL-6 KO mice exhibit only mild arthritis and the cartilage is histologically well-preserved. Similar to the observations made in the CIA model, both the antigen-specific proliferative response in lymph node cells and the in vivo antibody production elicited in IL-6 KO mice are reduced to less than half those seen in wild-type mice in the AIA model. Furthermore, the lymph node cells of IL-6 KO mice produce much more Th2-type cytokines than do wild-type cells (Ohshima et al. 1998). These results indicate that IL-6 is critical for the development of CIA and AIA, and may play a role in the Th1/Th2 response in these models.

Synovial fibroblastic cells are reported to proliferate in response to IL-6 plus sIL-6Rα, both of which are found in the synovial fluid of RA patients (Mihara et al. 1995). In addition, nonimmunologically mediated zymosan-induced arthritis is similar in wild-type and IL-6 KO mice at an early phase of the disease, but only wild-type mice exhibit chronic synovitis. Therefore, it is also likely that IL-6 causes the propagation of joint inflammation, possibly independent of its role in immunity (de Hooge et al. 2000)

Very recently, Atsumi et al. reported an interesting phenotype of a gp130 knock-in mouse strain. In this strain, named $gp130^{F759/F759}$, a point mutation is introduced to disrupt the gp130-mediated SHP-2/ERK MAPK signaling in vivo (see the next section for a description of the signal transduction, and refer to the section entitled "The signal orchestration model" for the knock-in strain). These mice spontaneously develop pathological symptoms highly similar to RA: symmetrical joint swelling and rigidness, marked proliferation of the synovium with pannus formation and fibrin deposits, infiltration of inflammatory cells into the joints, and severe bone destruction. In addition, activated osteoclasts are observed at the site of bone erosion. Furthermore, autoantibodies such as anti-DNA antibodies and rheumatoid factor are increased in the arthritic $gp130^{F759/F759}$ animals. The incidence of the disease reaches 100‰ in $gp130^{F759/F759}$ mice older than 16 months of age, suggesting that the point mutation in gp130 is a principal causal factor for the development of the RA-like disease in the $gp130^{F759/F759}$ mice. Importantly, when this $gp130^{F759/F759}$ strain is crossed with a lymphocyte-null RAG-2 KO strain, the joint disease disappears. Thus, the joint disease in $gp130^{F759/F759}$ mice is totally lymphocyte dependent, supporting the view that RA is an autoimmune disease. To our knowledge, this is the first and only direct evidence that a point mutation of a cytokine receptor has the potential to induce autoimmune disease (Atsumi et al. 2002).

Signal transduction

In this section, the molecular events that initiate and transduce the intracellular signaling through gp130 are described in detail. IL-6/gp130 signaling is the one of best-studied cascades among those of cytokine receptors. In addition, the understanding of gp130 signaling has been advanced by the identification of factors that negatively control it. Furthermore, knock-in gene targeting technology has made it possible to clarify the in vivo roles of each signaling pathway through gp130, and revealed that cytokine signaling is a more complicated and delicate biological event than had previously been thought.

Binding to the receptor

Upon IL-6 binding, the IL-6, IL-6Rα, and gp130 molecules form a hexameric complex with a stoichiometry of 2:2:2 (Ward et al. 1994). Mutagenesis along with epitope mapping studies and X-ray crystallography revealed important regions in the IL-6 protein, named sites 1–4, for the receptor binding and signal transduction. Site 1 is important for IL-6's binding with IL-6Rα. IL-6 mutants modified at site 2 or 3 can bind to the receptor but cannot transmit intracellular signals. The results of X-ray crystallography of IL-6 have led to a predicted role for the region designated as site 4 to stabilize the architecture of the signaling complex in the interaction between two IL-6/IL-6Rα/gp130 trimers (Somers et al. 1997). A scenario for the IL-6/IL-6 receptor interaction cascade can be depicted as follows: (1) IL-6 binds to IL-6Rα through site 1 of IL-6, forming a heterodimer; (2) the binary IL-6/IL-6Rα complex contacts the gp130 CBM through site 2 of IL-6 and also forms contacts between the C-terminal cell surface domain of IL-6Rα and gp130, resulting in a trimolecular complex with a 1:1:1 stoichiometry, which is not yet able to generate a signal (Chow et al. 2001); (3) two trimolecular complexes are assembled together via IL-6 site 3 and the Ig-like domain of gp130 (Moritz et al. 1999; Chow et al. 2001), as well as through site 4, completing the hexameric complex, which is competent to generate intracellular signals (Somers et al. 1997).

Intracellular signal transduction pathways

As described in the section entitled "Molecular aspects of the IL-6 receptor", gp130 has no intrinsic kinase activity but contains Box-motifs in its cytoplasmic domain; these motifs are known to associate with Janus kinases (JAK), which are nonreceptor tyrosine kinases (Figs. 2 and 4). Upon the binding of IL-6 to IL-6Rα, gp130 is homodimerized (Murakami et al. 1993), leading to the formation of the hexameric signaling-competent complex (Fig. 4). It is thought that the receptor homodimerization brings the JAK kinases into close proximity, resulting in their transactivation of each other. The activated JAK kinases phosphorylate tyrosine residues in the cytoplasmic domain of gp130. The human gp130 cytoplasmic domain has six tyrosine residues (Fig. 2). The first cytoplasmic tyrosine residue (Y683 in human gp130) is not part of a currently known sequence motif and is less significantly phosphorylated. The second tyrosine, which is positioned at the 759th amino acid residue of human gp130, resides within the $Y^{759}STV$ sequence of human gp130. This sequence is analogous to a motif for the recruitment of a protein tyrosine phosphatase, SHP-2 (Songyang et al. 1993; De Souza et al. 2002). In fact, SHP-2 is tyrosine phosphorylated upon IL-6 treatment and Y759 is required for the gp130-mediated phosphorylation of SHP-2 (Fukada et al. 1996). The third to sixth tyrosine residues (Y767, Y814, Y905, and Y915 in human gp130) form YXXQ motifs; this motif is responsible for the activation of a transcription factor, STAT (Stahl et al. 1995; Fukada et al. 1996; Gerhartz et al. 1996; Yamanaka et al. 1996; Figs. 2 and 4 . On the other hand, the cytoplasmic region of mouse gp130 contains seven tyrosine residues. When the mouse sequence is aligned with that of human gp130, the additional tyrosine (Y859 in mouse gp130) is found to be positioned between the *third* and *fourth* intracellular tyrosine residues (Fig. 2). The role of the additional tyrosine residue in gp130 signaling in the mouse is currently unknown. The remaining six tyrosine residues in mouse gp130 are completely analogous to their counterparts in the human protein: the second tyrosine, Y757, forms a YSTV sequence and quadruple YXXQ motifs are found for Y765, Y812, Y904, and Y914. In this review, unless otherwise noted, we will refer to the amino acid positions of human gp130, in particular for the second tyrosine, Y759. In brief, IL-6-induced, gp130-mediated JAK activation triggers two main signaling pathways: a Y759-derived SHP-2/ERK MAPK

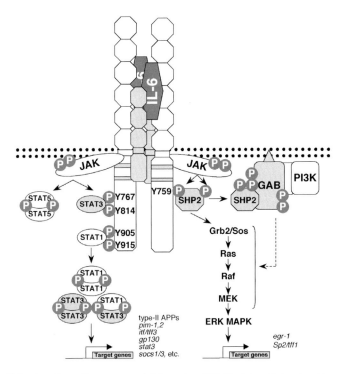

Fig. 4 Intracellular signaling pathways generated through gp130. The intracellular signal transduction pathways generated by IL-6 are shown. The formation of the hexameric IL-6/IL-6Rα/gp130 complex initiates signal transduction by activating JAK kinases. Phosphorylated tyrosine residues (P) in gp130 are recognized by SHP-2 and STAT molecules, leading to generation of the two major gp130 signaling pathways: the Y759-derived SHP-2/ERK MAPK cascade (*right side*) and the YXXQ-mediated STAT pathway (*left side*)

pathway and YXXQ-mediated STAT activation (Fig. 4; reviewed by Hirano et al. 1997; Hirano et al. 2000).

There are four members of the JAK family kinases found in mammals. Among them, gp130 constitutively associates with JAK1, JAK2, and TYK2 (Lutticken et al. 1994; Narazaki et al. 1994; Stahl et al. 1994). Several cell types isolated from JAK1 KO mice show diminished but still detectable DNA-binding activity of STAT3. However, JAK1 KO cells fail to exhibit biological responses induced by IL-6 family cytokines (Rodig et al. 1998). In contrast, fibroblasts from JAK2 KO mice show STAT3 phosphorylation and downstream gene transcription in response to IL-6 and sIL-6Rα (Parganas et al. 1998). Bone marrow-derived macrophages and embryonic fibroblasts from TYK2 KO mice also exhibit a STAT3 activation that is comparable to that of wild-type cells (Karaghiosoff et al. 2000; Shimoda et al. 2000). Similarly, when stimulated with IL-6 and sIL-6Rα, a great impairment of SHP-2 phosphorylation is observed for a fibrosarcoma cell line deficient for JAK1, but not JAK2 or TYK2 (Schaper et al. 1998). These results suggest that, among the JAK kinases associated with gp130, JAK1 serves a major role in the gp130-mediated SHP-2 and STAT3 pathways.

IL-6 signal transduction has been extensively investigated using cell lines stably transfected with a chimeric gp130 molecule, consisting of the extracellular domain of G-CSF

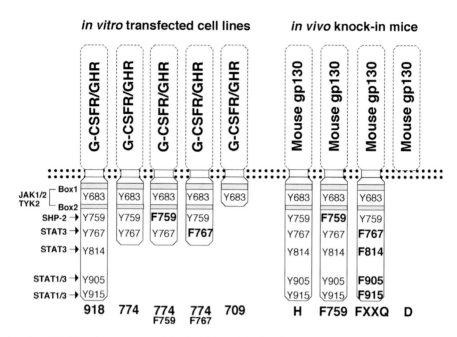

Fig. 5 gp130 Chimeric receptors used for elucidation of the signaling pathways. The gp130 chimeric receptors used for in vitro (*left*) and in vivo (*right*) studies are illustrated. For the in vitro studies, intracellular human gp130 (full length: 918) or truncated mutants with or without a Y to F substitution (709, 774, 774/F759 and 774/F767) is fused with the extracellular domain of the receptor for rabbit GH or human G-CSF. To generate knock-in mouse strains, chimeric receptors consisting of intracellular human gp130 and extracellular mouse gp130 domains are used. In vivo, the replacement of Y759 with F (*F759*) also leads to a defect in the Y759-derived SHP-2/ERK MAPK cascade, mutations in all four YXXQ motifs (*FXXQ*) abrogate the YXXQ-mediated STAT activation, and a loss of the cytoplasmic domain of gp130 (*D*) abolishes the generation of signals through gp130. The FXXQ and D knock-in strains are lethal within 24 h after birth

or GH connected with the intracellular region of human gp130 (Fig. 5). The chimeric receptor system is advantageous when gp130 is endogenously expressed on the cells to be analyzed. By replacing the intracellular tyrosine(s) with phenylalanine or by constructing truncated mutants of the gp130 tail, the signaling cascades and physiological functions carried by the respective tyrosine residues in the gp130 cytoplasmic domain have now been elucidated (see next sections).

The gp130 Y759-mediated SHP-2/ERK MAPK pathway

The amino acid sequence surrounding Y759 in human gp130, the second tyrosine from the plasma membrane, comprises a context similar to the SHP-2-binding motif (Songyang et al. 1993; De Souza et al. 2002; Fig. 2). In fact, an oligopeptide representing amino acid residues 752–766 of gp130 containing phosphorylated, but not nonphosphorylated Y759, can bind to and coprecipitate with SHP-2 (Fukada et al. 1996). SHP-2 is a protein tyrosine phosphatase that also contains two tandem SH2 domains in its N-terminal region, which are known to bind to phosphotyrosine residues; it also has several potential Grb2 (growth factor receptor-bound protein 2)-binding motifs in its C-terminal region (Feng et al. 1993;

Vogel et al. 1993). The SHP-2 phosphatase activity has been shown to be important for growth factor-mediated ERK MAPK activation (Bennett et al. 1996). In addition, hepatoma cells that overexpress a truncated form of a SHP-2 mutant that cannot bind to Grb2 exhibit a substantial decrease in IL-6-induced ERK MAPK activation (Kim and Baumann 1999). Thus, SHP-2 is involved in the ERK MAPK cascade by acting both as an enzyme and as an adapter protein. The phosphatase activity of SHP-2 is regulated by its SH2 domain: in the free, unbound state of SHP-2, the N-terminal SH2 domain interacts with its own phosphatase domain, blocking the active site. The binding of the SH2 domain to a phosphotyrosine residue unblocks the active site, activating the protein's phosphatase activity (Hof et al. 1998). As illustrated in Fig. 4, upon IL-6 stimulation, SHP-2 is recruited to the phosphorylated Y759 residue of gp130. After being recruited, SHP-2 is phosphorylated by JAKs and then interacts with Grb2, which is constitutively associated with Sos (son-of-sevenless), a GDT/GTP exchanger for Ras. The GTP form of Ras transmits signals that lead to activation of the ERK MAPK cascade. The ERK MAPK activation is not observed in cells lacking the SHP-2-binding site (Y759) of gp130 (Fukada et al. 1996; Ohtani et al. 2000; Tebbutt et al. 2002). Furthermore, a gp130 mutant lacking all the cytoplasmic tyrosine residues except Y759 associates with SHP-2 at a level comparable to wild-type intact gp130 (Anhuf et al. 2000). These results indicate that Y759 is necessary and sufficient for the gp130-mediated SHP-2/ERK MAPK cascade.

In our studies, we found that after gp130 stimulation, a tyrosine-phosphorylated protein with a molecular mass of approximately 100 kDa –110 kDa was coprecipitated with SHP-2 (Fukada et al. 1996). The 100 kDa –110 kDa protein was identified as Gab1 (Grb2-associating binder 1), based on its recognition by an antibody raised against Gab1 (Takahashi-Tezuka et al. 1998). The Gab family proteins are adaptor/scaffold proteins containing several tyrosine residues that undergo phosphorylation and multiple functional motifs. These motifs include a PH (pleckstrin homology) domain, proline-rich domain, c-Met-binding domain (at least in Gab1), and docking sites for Grb2, Crk/PLCγ, SHP-2, and the p85 subunit of PI3 K (phosphatidylinositol 3'-kinase; reviewed by Hibi and Hirano 2000; Liu and Rohrschneider 2002). To date, Gab1–3 have been reported in humans and mice (Holgado-Madruga et al. 1996; Nishida et al. 1999; Wolf et al. 2002). At least Gab1 and Gab2 undergo tyrosine-phosphorylation, most likely by JAKs, in a gp130 stimulation-dependent manner. Because the phosphorylation of Gab1 and Gab2 is still observed in cells transfected with a gp130 mutant lacking all six cytoplasmic tyrosine residues or a gp130[709] truncation mutant (which carries Box-1 motif and the first tyrosine; see Fig. 5), the phosphorylation of Gab proteins occurs independent of the tyrosine-phosphorylation state of gp130 (Takahashi-Tezuka et al. 1998; Nishida et al. 1999). The tyrosine-phosphorylated Gab1 and Gab2 interact with SHP-2 and the p85 subunit of PI3K (Fig. 4). The tertiary complex of SHP-2/Gab1/PI3K is not observed when Y759 in gp130 is mutated, indicating that Gab1-mediated signaling is dependent on Y759 of gp130. The formation of the tertiary complex can lead to activation of the Ras/Raf/ERK MAPK cascade, which is based on the observations that the forced expression of Gab1 or Gab2 protein enhances the ERK MAPK activation, and the inhibition of PI3K by wortmannin or a dominant-negative form of the p85 subunit decreases it (Takahashi-Tezuka et al. 1998; Nishida et al. 1999). Moreover, in embryonic fibroblasts isolated from Gab1 KO mice, the activity of the ERK MAPK pathway upon stimulation with IL-6/IL-6Rα was severely impaired (Itoh et al. 2000), demonstrating a requirement for Gab1 in the gp130 Y759-mediated ERK MAPK pathway in fibroblasts.

One of the target genes of the gp130 Y759-signaling pathway is a zinc finger transcription factor, *egr-1*. The expression of *egr-1* is induced by IL-6 in the mouse leukemia cell line, M1. Mutation at Y759 in gp130 abrogates the *egr-1* mRNA expression in M1 transfectants (Yamanaka et al. 1996). In addition, a pharmacological inhibition of the ERK MAPK pathway by PD98059 diminishes IL-6-induced *egr-1* transcription in a hepatoma cell line (Kim and Baumann 1999). The promoter region of the *egr-1* gene contains several serum-response elements (SRE), suggesting that gp130 Y759-mediated ERK MAPK activation induces the phosphorylation of Elk-1, a binding partner for serum-responsive factor, driving the SRE-dependent *egr-1* expression. Another gene recently reported to be targeted by the gp130-mediated SHP-2/ERK MAPK cascade is the gene encoding Sp2/TFF1 (trefoil factor-1), which contains an AP-1 site in its promoter region. A mutation in the SHP-2 binding site of murine gp130 diminished the activation of the *Sp2/tff1* promoter in vitro and significantly reduced the level of Sp2/TFF1 protein in vivo (in the stomach), indicating that gastric Sp2/TFF1 is a direct target of the gp130-mediated SHP-2/ERK MAPK pathway (Tebbutt et al. 2002).

The gp130 YXXQ-mediated JAK/STAT pathway

Quadruple STAT-binding YXXQ motifs are found in the cytoplasmic domain of gp130 (Figs. 2 and 4). STAT proteins are transcription factors that dimerize upon being phosphorylated on tyrosine by JAK, after which they enter the nucleus and transactivate target genes. So far, seven STAT proteins (STAT1 to 4, 5a, 5b, and 6) have been identified in humans and mice (reviewed by Bromberg and Darnell 2000; O'Shea et al. 2002). It is reported that gp130 activation by IL-6 induces the activation of STAT1, 3, and 5. For STAT3 activation, any one of the four YXXQ motifs is sufficient, while the distal two YXPQ sequences are important for STAT1 activation (Gerhartz et al. 1996; Figs. 2 and 4). Although there is a redundancy of YXXQ motifs for STAT3 activation, Schmitz et al. showed that the four STAT binding sites in gp130 are not equivalent: the distal two tyrosines (Y905 or Y915), which form YXPQ motifs, provide more potent STAT activation, in terms of the DNA-binding activity of STAT3 and APP gene promoter activation, than do the proximal YXXQ motifs (Y767 or Y814; Schmitz et al. 2000a). Upon stimulation by IL-6, STAT proteins are recruited to the phosphorylated YXXQ/YXPQ motifs and then phosphorylated by JAKs. The activated STAT proteins form a heterodimer (STAT1:STAT3) or homodimers (STAT1:STAT1 and/or STAT3:STAT3), subsequently translocate to the nucleus and activate the transcription of target genes. STAT3 is also phosphorylated on serine residues by an H7-sensitive kinase pathway. This phosphorylation is necessary for the full transcriptional activity of STAT3 (Abe et al. 2001).

The gp130-mediated STAT3 activation induces the expression of many genes, including genes encoding the type-II APPs (reviewed by Moshage 1997), *pim-1* and *pim-2* proto-oncogenes (Shirogane et al. 1999), *itf/tff3* (Tebbutt et al. 2002), and negative regulators for the JAK/STAT pathways, *socs1/jab/ssi1* (Endo et al. 1997; Naka et al. 1997; Starr et al. 1997) and *socs3/jab2/cis3/ssi2* (Schmitz et al. 2000b). Interestingly, the *gp130* (O'Brien and Manolagas 1997) and *stat3* (Ichiba et al. 1998; Narimatsu et al. 2001) genes are also direct targets of gp130-mediated STAT signaling, suggesting the existence of an autoregulatory mechanism for this signaling pathway.

Turn-off signals

After gp130 stimulation, the phosphorylation of ERK MAPK and STAT3 is rapidly induced, and then gradually attenuated in many cell types, such as primary fibroblasts. This suggests the existence of a cellular mechanism that turns off the signaling. The mechanism underlying the downregulation or attenuation of the gp130 signal has been largely revealed by identification of the molecules that operate it. One is TC45, the nuclear isoform of TC-PTP (T-cell protein tyrosine phosphatase), which was recently shown to dephosphorylate STAT proteins. In TC-PTP null embryonic fibroblasts, the dephosphorylation of IL-6-activated STAT3 is impaired (ten Hoeve et al. 2002). Another molecule that interacts with STAT1, named PIAS1 (protein inhibitor of activated STAT1), was identified by the yeast two-hybrid method. Searching the EST database and screening cDNA libraries has led to the identification of several proteins related to PIAS1, including PIAS3. PIAS1 and PIAS3 specifically associate with ligand-stimulated, i.e., tyrosine-phosphorylated, STAT1 and STAT3, respectively. PIAS1 blocks the DNA-binding activity of STAT and inhibits STAT1-mediated gene induction, but not STAT3 activity. Conversely, PIAS3 specifically represses STAT3 activity (Chung et al. 1997; Liu et al. 1998). Similarly, the SOCS family proteins are also negative regulators of IL-6 signaling. SOCS1, also known as JAB and SSI1, can mask a critical tyrosine residue within the activation loop of the JAK kinase domain, thereby inhibiting JAK activity (reviewed by Yasukawa et al. 2000). On the other hand, SOCS3, also known as CIS3, JAB2, and SSI3, is expressed by IL-6 stimulation and binds to phosphorylated Y759 in gp130, leading to attenuation of the gp130 signaling (Nicholson et al. 2000; Schmitz et al. 2000b). The phosphorylation state of gp130, JAK, and STAT3 is prolonged in the primary fibroblasts from a Y759-lacking knock-in mouse, $gp130^{F759/F759}$ (see the section entitled "The signal orchestration model" for a description of the knock-in strain) and certain cell lines carrying the gp130 Y759F mutation (Kim et al. 1998; Schaper et al. 1998; Ohtani et al. 2000), which is partly explained by the loss of the SOCS3-binding site, i.e., Y759, in gp130. Because SHP-2 is also recruited to the phosphorylated Y759 of gp130, however, the negative regulatory effect of its tyrosine phosphatase activity on gp130 signaling can also be considered. A catalytically inactive mutant of SHP-2 enhances the activation of the APP gene promoter in response to gp130 stimulation, but this enhancement is not observed when Y759 in gp130 is mutated (Kim et al. 1998; Schaper et al. 1998; Kim and Baumann 1999). A recent report by Lehmann et al. showing that both SHP-2 and SOCS3 contribute to the Y759-dependent attenuation of IL-6/gp130 signaling demonstrated this regulation more clearly (Lehmann et al. 2002). Finally, as described in the section entitled "Molecular aspects of the IL-6 receptor", the antagonizing effect of sgp130 could also contribute to the turning off of gp130 signaling.

The signal orchestration model

We have performed studies using cell lines transfected with chimeric receptors consisting of the extracellular domain of GH or G-CSF fused with the intracellular region of gp130 (Fig. 5). Both GH and G-CSF mediate the homodimerization of these chimeric receptors, therefore mimicking the IL-6-induced homodimerization and activation of gp130. These systems have greatly contributed to unraveling the signal transduction pathways and physiological roles carried by each tyrosine residue of gp130. Using the in vitro chimeric receptor system, we found that gp130 can simultaneously activate contradictory signals in a given target cell. In this subsection, we discuss this finding by presenting some examples.

The first example comes from studies using a murine myeloid leukemic M1 cell line, which shows growth arrest and differentiation into a macrophage-like morphology in response to gp130 stimulation. M1 cells carrying the GHR/gp130[774] chimera (see Fig. 5), which is a gp130-truncated mutant that contains the Box motifs, SHP-2-binding site (Y759), and one membrane-proximal YXXQ motif (Y767), is still able to differentiate into a macrophage-like shape in response to the chimeric receptor ligand, GH, suggesting a redundancy of the four STAT-binding sites in gp130. The requirement of STAT3 activation for the differentiation of M1 cells was demonstrated by the findings that mutation of the YXXQ motif to FXXQ, as in GHR/gp130[774/F767], renders the M1 transfectants incapable of differentiation (Yamanaka et al. 1996), and a dominant-negative form of STAT3 inhibits the differentiation of parental M1 cells (Nakajima et al. 1996). On the other hand, the GHR/gp130[774/F767]-expressing M1 transfectants, which are defective in STAT3 activation but have intact SHP-2 signaling, show enhanced cell growth, suggesting that growth-enhancing signal(s), opposite to the YXXQ-mediated ones responsible for growth arrest, are derived from Y759 of gp130 (Nakajima et al. 1996).

Likewise, the G-CSFR/gp130 chimera was introduced into the IL-3-dependent pro-B cell line, BAF/B03, which proliferates in response to gp130 signaling stimulated by the chimeric receptor ligand, G-CSF. BAF/B03 cells carrying the G-CSFR/gp130[774]-truncated chimera grow comparably to cells expressing the chimera with the full-length, wild-type cytoplasmic region, G-CSFR/gp130[918]. Mutation at Y759 (G-CSFR/gp130[774/F759]) causes a pause in cell proliferation, but does not cause cell death for up to 4 d in culture. In contrast, FXXQ mutation (such as in G-CSFR/gp130[774/F767]) or the introduction of a dominant-negative form of STAT3 leads to apoptosis of the transfectants in the first 24 h of culture, suggesting that STAT3 activation is essential for antiapoptotic signals in this cell line. Upon stimulation, BAF/B03 cells expressing the G-CSFR/gp130[774/F759] mutant fail to make the cell-cycle transition from the S to the G2/M phase, indicating that Y759 signaling mediates cell-cycle progression (Fukada et al. 1996). A cDNA subtraction experiment revealed that the proto-oncogenes *pim-1* and *pim-2* are among the targets of the gp130-mediated STAT3 activation, and are required for the gp130-mediated antiapoptotic activity and G1 to S cell-cycle transition (Shirogane et al. 1999). Thus, gp130-mediated STAT3 signaling is positively involved in the G1 to S cell-cycle transition in BAF/B03 cells. However, under conditions when gp130-mediated STAT3 signaling is suppressed, gp130 induces the upregulation of cyclin-dependent kinase inhibitor, p21, in BAF/B03 cells. This is the second example showing that gp130 can simultaneously activate contradictory signals in a given target cell (Fukada et al. 1998).

The third example is observed in the rat pheochromocytoma PC12 cell line, which extends neurites in response to IL-6 when pretreated with nerve-growth factor. In these cells, gp130-derived, SHP-2-mediated ERK MAPK signals positively control the neurite outgrowth, and the YXXQ-mediated STAT3 activation negates the effect of gp130 stimulation.

These in vitro transfectant studies provide evidence that distinct intracellular signaling pathways generated by a given ligand can carry different, sometimes opposite physiological role(s). The overall balance of these distinct signals could determine the final biological outputs by a given ligand in a target cell (Fig. 6). We have proposed to call this concept *the signal orchestration model* (Hirano et al. 1997; Hirano 1999; Hirano and Fukada 2001). The simultaneous generation of contradictory signals is not unique to gp130 signal transduction. For example, the binding of TNFα to TNFR1 induces receptor trimerization

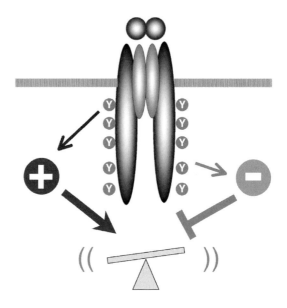

Fig. 6 The signal orchestration model. A ligand can simultaneously induce contradictory signals in a given target cell through distinct regions of its receptor. The balance of the contradictory signals elicited by a receptor (depicted as a *seesaw*) determines the final output of biological activity of a given ligand in a given target cell

and activates a caspase cascade. Initiation of the caspase cascade results in the apoptosis of a target cell. However, signal transduction via TNFR1 also involves the activation of transcription factors, AP-1 and NF-κB, which subsequently leads to the generation of anti-apoptotic and inflammatory responses (reviewed in Baud and Karin 2001). Thus, the orchestration of TNFα/TNFR1 signaling could determine the final output, i.e., apoptosis or inflammatory responses. Among the molecules modulating the TNFα/TNFR1 signal orchestration are the cIAPs (cellular inhibitors of apoptosis). TNFα-induced, NF-κB-mediated induction of cIAPs counteracts the TNFα-induced apoptosis through the inhibition of caspase activation (reviewed in Baud and Karin 2001). In another example, both cyclin D1 and a cyclin-dependent kinase inhibitor, p21, can be activated by growth factors (Schreiber et al. 1999). Taking these observations into consideration, certain cytokines and growth factors have the potential to trigger contradictory signals within a target cell, simultaneously. Under physiological conditions, cells would be exposed to a milieu containing a cytokine cocktail. Hence, in this model, cross-talk among different signaling networks can influence the level or activation state of *the conductor* of a certain signal, which modulates the balance of the signaling and determines the biological outputs. This mechanism may explain how a pleiotropic cytokine like IL-6 can exert multiple functions in a variety of cells.

The evidence discussed so far in this subsection is based on results from in vitro studies. To address the in vivo roles of each signal transduction pathway through gp130, we generated a series of knock-in mouse strains expressing various gp130 chimeras (extracellular domain of mouse gp130/cytoplasmic region of human gp130; Fig. 5). Neonates of a knock-in mouse strain carrying a gp130 mutant with the whole cytoplasmic domain deleted ($gp130^{D/D}$), and a STAT signaling-defective gp130 mutant ($gp130^{FXXQ/FXXQ}$) die within 24 h after birth, demonstrating a vital role for gp130-mediated STAT activation in postnatal life. To investigate the function of gp130 YXXQ-mediated signaling in the immune system, fetal liver cells (FL) obtained from $gp130^{FXXQ/FXXQ}$ or $gp130^{D/D}$ embryos were transferred into lethally X-ray-irradiated wild-type mice to obtain FL-reconstituted mice, in which the hematopoietic cells are of the mutant origin (fetal liver chimera). Serum Ig levels are decreased in $gp130^{FXXQ/FXXQ}$ or $gp130^{D/D}$ FL-reconstituted mice, as compared with wild-type FL-reconstituted mice. Antigen-specific antibody production by immunization with a thymus-dependent antigen (the TD response) is also suppressed in the $gp130^{FXXQ/FXXQ}$ FL-reconstituted mice, indicating that gp130-mediated STAT activation is critical for Ig production. In contrast, mice from another knock-in stain carrying a point mutation in the SHP-2-binding site ($gp130^{F759/F759}$), thus defective in the SHP-2/ERK MAPK cascade, are born apparently healthy and are fertile. However, as they aged, the $gp130^{F759/F759}$ mice exhibit splenomegaly, lymphadenopathy, and increased serum Ig levels and the TD response, indicating a negative regulatory role for Y759 in leukocyte numbers and Ig production. When primary fibroblasts obtained from the $gp130^{F759/F759}$ mice are stimulated with IL-6/IL-6Rα, the duration of the phosphorylation states of gp130, JAK, and STAT3 were prolonged compared with the duration in wild-type fibroblasts, indicating the mechanism underlying the negative control effect of Y759 (Ohtani et al. 2000). As described in the section entitled "Relevance of the IL-6 to human diseases", these phenotypic changes are followed by the development of an RA-like joint disease. A possible mechanism for the disease development in $gp130^{F759/F759}$ mice was indicated by Atsumi et al., who showed that IL-6 can inhibit the activation-induced cell death (AICD) of T cells in vitro and that the inhibitory effect is more pronounced in T cells from $gp130^{F759/F759}$ mice than from wild-type mice. Because the T cells from $gp130^{FXXQ/FXXQ}$ FL-reconstituted mice efficiently undergo AICD even in the presence of IL-6, the inhibitory effect of IL-6 is probably dependent on STAT activation. In addition, the stronger inhibition of AICD in $gp130^{F759/F759}$ T cells is correlated with prolonged STAT3 activation as a consequence of the signaling imbalance by the Y759F point mutation. Because AICD is a process that is known to eliminate self-reactive T cells, the defective AICD in $gp130^{F759/F759}$ T cells is one of the causal factors for the breakdown of self-tolerance, possibly leading to the development of the RA-like disease (Atsumi et al. 2002). These observations demonstrate a negative regulatory role for Y759-mediated signals in vivo. When $gp130^{F759/F759}$ mice are infected in vivo with gram-positive bacteria, *Listeria monocytogenesis*, however, they die more rapidly and with a higher bacterial burden than do wild-type controls (Kamimura et al. 2002), suggesting that the positive effect of Y759 signaling is generated as a defense mechanism during the early phase of bacterial infection.

Another set of gp130 knock-in mouse strains, designated $gp130^{\Delta STAT}$ and $gp130^{757F}$ (where the number represents the amino acid position of mouse gp130) has recently been generated by another group. One of the knock-in strains, termed $gp130^{\Delta STAT}$ mice, harbors a truncated form of gp130 containing the SHP-2-binding site and a mutated STAT3-binding site. This mutation causes inactivation of the gp130-mediated STAT responses, similar to gp130[774/F767] (Fig. 5). Surprisingly, despite the defective gp130-mediated STAT3 activation, this stain of mice survives for at least 6–8 months (Table 2). Moreover, these mice spontaneously develop gastrointestinal ulceration and degenerative joint disease accompanied by the synovial hyperplasia. As a consequence of the gp130-mediated STAT-signaling defect, a sustained phosphorylation of JAK2, SHP-2, and ERK MAPK in response to IL-6 and sIL-6Rα is observed in the synovial cells of the $gp130^{\Delta STAT}$ mice. Because SOCS1-deficient synovial fibroblasts also show an extended period of ERK MAPK activation when stimulated with IL-6 and sIL-6Rα, a loss of gp130 STAT-mediated SOCS1 expression is thought to be one mechanism for the sustained phosphorylation of these molecules and a possible cause of the synovial hyperplasia in the $gp130^{\Delta STAT}$ mice

(Ernst et al. 2001). Another knock-in strain, named $gp130^{757F}$, was designed to be incapable of activating the gp130/SHP-2 ERK MAPK pathway, theoretically similar to our $gp130^{F759/F759}$ mice. The $gp130^{757F}$ mice exhibit splenomegaly and spontaneously develop gastric adenoma 6–8 weeks after birth. The authors noticed that this phenotype is essentially phenocopied by pS2/TFF1 (trefoil factor-1) KO mice (Lefebvre et al. 1996). TFF comprises a family that includes pS2/TFF1, TFF2 and ITF/TFF3, which are expressed in the stomach, stomach/pancreas, and intestine/colon, respectively. These proteins are known to have cytoprotective effects that promote wound healing in response to gastrointestinal injury. The level of pS2/TFF1 protein is eventually decreased in the stomach of $gp130^{757F}$ mice and the promoter activity of the $pS2/tff1$ gene is regulated by the gp130/SHP-2 ERK MAPK pathway. In a reciprocal fashion, the level of ITF/TFF3 protein is reduced in the colon of $gp130^{\Delta STAT}$ mice and its expression is suggested to be directly regulated by gp130-mediated STAT3 activation. In line with this, $gp130^{\Delta STAT}$ mice are highly sensitive to sodium dextran sulfate-induced experimental colitis, compared with their wild-type littermates. In contrast, $gp130^{757F}$ mice are completely resistant to the colitis, which is associated with the increase of ITF/TFF3 protein level in the intestine. Thus, gp130 reciprocally regulates gastrointestinal homeostasis by inducing tissue-specific TFFs through the gp130-mediated SHP-2/ERK MAPK or STAT3 pathway (Tebbutt et al. 2002).

These in vivo studies provide further evidence that the ablation of one signaling cascade can influence the other—that is, a loss of Y759-mediated signaling enhances the YXXQ-mediated STAT activation, and vice versa. However, through this mechanism a risk arises for unexpected biological consequences (Fig. 7). The difference in phenotypes between our $gp130^{F759/F759}$ mice and the $gp130^{757F}$ mice of Ernst's group is intriguing. Both strains of mice are expected to have specific disruption of the gp130-mediated SHP-2/ERK MAPK cascade in vivo due to point mutation(s). Splenomegaly is a common phenotype to both these mutants. However, we did not clearly recognize enlargement of the stomach and proximal small intestine in $gp130^{F759/F759}$ mice. In contrast, no joint pathology was documented for the $gp130^{757F}$ mice (Tebbutt et al. 2002). The following possibilities for this phenotypic discrepancy can be raised: (a) differences in environmental factors for the mice, including the feed and rearing space used, (b) dissimilarity of the microflora making up the commensal bacteria, and (c) a difference in targeting strategy for the knock-in strains. To destroy the gp130-mediated SHP-2/ERK MAPK cascade in vivo, we used a cDNA for the intracellular domain of *human* gp130 and introduced a single point mutation of Y759F in the $gp130^{F759/F759}$ knock-in mice, while Ernst's group mutated the endogenous (i.e., *mouse*) gp130 by replacing two amino acid residues (Y757F and V760A) in the $gp130^{757F}$ knock-in mice. Such a difference in targeting strategy might influence the finely tuned endogenous gp130 signal. For example, if mouse STAT3 interacts with the YXXQ motifs of human gp130 with a somewhat different affinity than with the endogenous mouse YXXQ motifs, $gp130^{F759/F759}$ mice bearing the human gp130 cytoplasmic domain might exhibit an altered level of STAT3 activation compared with the endogenous gp130/STAT3 interaction. The phenotypic inconsistency between the $gp130^{F759/F759}$ and $gp130^{757F}$ mice also implies that the modulation of gp130 signaling does not occur in an all-or-none fashion. Rather, gp130 could exhibit various biological functions in a manner that depends on the delicate tuning of its signaling. A different degree of disturbance in *the signal orchestration* through gp130 may produce these dramatic phenotype differences between the $gp130^{F759/F759}$ and the $gp130^{757F}$ mice.

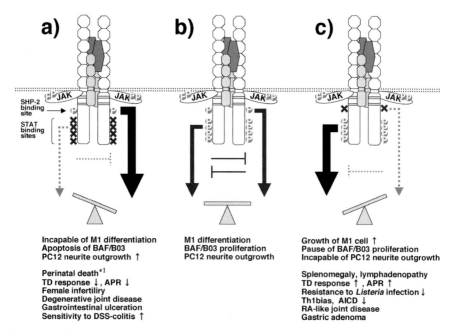

Fig. 7a–c Consequences of disturbance in the gp130 signal orchestration. A ligand binding to gp130 initiates two major signal transduction pathways, as illustrated in Fig. 4. For convenience, gp130-mediated STAT signaling is depicted on the *left* side and the SHP-2/ERK MAPK cascade is on the *right* side of the receptor. Both signaling pathways mutually regulate each other. **a** A defect in STAT activation leads to enhancement of the signaling derived from the SHP-2 binding site. **b** Intact gp130 generates *neutral* signaling. **c** A mutation at the SHP-2-binding site results in sustained STAT3 activation. The overall balance of gp130 signaling is represented by a *seesaw*. The physiological consequences of a disturbance in the signal orchestration are listed below the seesaw. *1; in the case of $gp130^{FXXQ/FXXQ}$

Taking the results of these in vitro and in vivo studies together, IL-6 can generate diverse gp130-derived signaling pathways, which mainly consist of the Y759-mediated SHP-2/ERK MAPK cascade and YXXQ-derived STAT activation. These intracellular signals carry respective biological effects by activating different sets of genes, but the two signaling cascades are not mutually exclusive, since the regulation of both pathways is reciprocally dependent (Figs. 6 and 7).

References

Abe K, Hirai M, Mizuno K, Higashi N, Sekimoto T, Miki T, Hirano T, Nakajima K (2001) The YXXQ motif in gp 130 is crucial for STAT3 phosphorylation at Ser727 through an H7-sensitive kinase pathway. Oncogene 20:3464–3474

Alexander WS, Rakar S, Robb L, Farley A, Willson TA, Zhang JG, Hartley L, Kikuchi Y, Kojima T, Nomura H, Hasegawa M, Maeda M, Fabri L, Jachno K, Nash A, Metcalf D, Nicola NA, Hilton DJ (1999) Suckling defect in mice lacking the soluble haemopoietin receptor NR6. Curr Biol 9:605–608

Alonzi T, Fattori E, Lazzaro D, Costa P, Probert L, Kollias G, De Benedetti F, Poli V, Ciliberto G (1998) Interleukin 6 is required for the development of collagen-induced arthritis. J Exp Med 187:461–468

Anhuf D, Weissenbach M, Schmitz J, Sobota R, Hermanns HM, Radtke S, Linnemann S, Behrmann I, Heinrich PC, Schaper F (2000) Signal transduction of IL-6, leukemia-inhibitory factor, and oncostatin M: structural receptor requirements for signal attenuation. J Immunol 165:2535–2543

Astaldi GC, Janssen MC, Lansdorp P, Willems C, Zeijlemaker WP, Oosterhof F (1980) Human endothelial culture supernatant (HECS): a growth factor for hybridomas. J Immunol 125:1411–1414

Atreya R, Mudter J, Finotto S, Mullberg J, Jostock T, Wirtz S, Schutz M, Bartsch B, Holtmann M, Becker C, Strand D, Czaja J, Schlaak JF, Lehr HA, Autschbach F, Schurmann G, Nishimoto N, Yoshizaki K, Ito H, Kishimoto T, Galle PR, Rose-John S, Neurath MF (2000) Blockade of interleukin 6 trans signaling suppresses T-cell resistance against apoptosis in chronic intestinal inflammation: evidence in Crohn's disease and experimental colitis in vivo. Nat Med 6:583–588

Atsumi T, Ishihara K, Kamimura D, Ikushima H, Ohtani T, Hirota S, Kobayashi H, Park SJ, Saeki Y, Kitamura Y, Hirano T (2002) A point mutation of Tyr-759 in interleukin 6 family cytokine receptor subunit gp130 causes autoimmune arthritis. J Exp Med 196:979–990

Baud V, Karin M (2001) Signal transduction by tumor necrosis factor and its relatives. Trends Cell Biol 11:372–377

Baumann H, Gauldie J (1994) The acute phase response. Immunol Today 15:74–80

Bennett AM, Hausdorff SF, O'Reilly AM, Freeman RM, Neel BG (1996) Multiple requirements for SH-PTP2 in epidermal growth factor-mediated cell cycle progression. Mol Cell Biol 16:1189–1202

Bernad A, Kopf M, Kulbacki R, Weich N, Koehler G, Gutierrez-Ramos JC (1994) Interleukin-6 is required in vivo for the regulation of stem cells and committed progenitors of the hematopoietic system. Immunity 1:725–731

Bethin KE, Vogt SK, Muglia LJ (2000) Interleukin-6 is an essential, corticotropin-releasing hormone- independent stimulator of the adrenal axis during immune system activation. Proc Natl Acad Sci USA 97:9317–9322

Betz UA, Bloch W, van den Broek M, Yoshida K, Taga T, Kishimoto T, Addicks K, Rajewsky K, Muller W (1998) Postnatally-induced inactivation of gp130 in mice results in neurological, cardiac, hematopoietic, immunological, hepatic, and pulmonary defects. J Exp Med 188:1955–1965

Borsellino N, Bonavida B, Ciliberto G, Toniatti C, Travali S, D'Alessandro N (1999) Blocking signaling through the Gp130 receptor chain by interleukin-6 and oncostatin M inhibits PC-3 cell growth and sensitizes the tumor cells to etoposide and cisplatin-mediated cytotoxicity. Cancer 85:134–144

Bowcock AM, Kidd JR, Lathrop GM, Daneshvar L, May LT, Ray A, Sehgal PB, Kidd KK, Cavalli-Sforza LL (1988) The human "interferon-β2/hepatocyte stimulating factor/interleukin- 6" gene: DNA polymorphism studies and localization to chromosome 7p21. Genomics 3:8–16

Bravo J, Heath JK (2000) Receptor recognition by gp130 cytokines. EMBO J 19:2399–2411

Bromberg J, Darnell JE, Jr. (2000) The role of STATs in transcriptional control and their impact on cellular function. Oncogene 19:2468–2473

Brown S, Hu N, Hombria JC (2001) Identification of the first invertebrate interleukin JAK/STAT receptor, the Drosophila gene domeless. Curr Biol 11:1700–1705

Brunello AG, Weissenberger J, Kappeler A, Vallan C, Peters M, Rose-John S, Weis J (2000) Astrocytic alterations in interleukin-6/Soluble interleukin-6 receptor αdouble-transgenic mice. Am J Pathol 157:1485–1493

Butzkueven H, Zhang JG, Soilu-Hanninen M, Hochrein H, Chionh F, Shipham KA, Emery B, Turnley AM, Petratos S, Ernst M, Bartlett PF, Kilpatrick TJ (2002) LIF receptor signaling limits immune-mediated demyelination by enhancing oligodendrocyte survival. Nat Med 8:613–619

Campbell IL, Abraham CR, Masliah E, Kemper P, Inglis JD, Oldstone MB, Mucke L (1993) Neurologic disease induced in transgenic mice by cerebral overexpression of interleukin 6. Proc Natl Acad Sci USA 90:10061–10065

Chen HW, Chen X, Oh SW, Marinissen MJ, Gutkind JS, Hou SX (2002) mom identifies a receptor for the Drosophila JAK/STAT signal transduction pathway and encodes a protein distantly related to the mammalian cytokine receptor family. Genes Dev 16:388–398

Chesnokova V, Auernhammer CJ, Melmed S (1998) Murine leukemia inhibitory factor gene disruption attenuates the hypothalamo-pituitary-adrenal axis stress response. Endocrinology 139:2209–2216

Chomarat P, Banchereau J, Davoust J, Palucka AK (2000) IL-6 switches the differentiation of monocytes from dendritic cells to macrophages. Nat Immunol 1:510–514

Chow D, He X, Snow AL, Rose-John S, Garcia KC (2001) Structure of an extracellular gp130 cytokine receptor signaling complex. Science 291:2150–2155

Christadoss P, Poussin M, Deng C (2000) Animal models of myasthenia gravis. Clin Immunol 94:75–87

Chung CD, Liao J, Liu B, Rao X, Jay P, Berta P, Shuai K (1997) Specific inhibition of Stat3 signal transduction by PIAS3. Science 278:1803–1805

Clegg CH, Haugen HS, Rulffes JT, Friend SL, Farr AG (1999) Oncostatin M transforms lymphoid tissue function in transgenic mice by stimulating lymph node T-cell development and thymus autoantibody production. Exp Hematol 27:712–725

Content J, De Wit L, Poupart P, Opdenakker G, Van Damme J, Billiau A (1985) Induction of a 26-kDa-protein mRNA in human cells treated with an interleukin-1-related, leukocyte-derived factor. Eur J Biochem 152:253–257

Dalrymple SA, Slattery R, Aud DM, Krishna M, Lucian LA, Murray R (1996) Interleukin-6 is required for a protective immune response to systemic *Escherichia coli* infection. Infect Immun 64:3231–3235

de Hooge AS, van De Loo FA, Arntz OJ, van Den Berg WB (2000) Involvement of IL-6, apart from its role in immunity, in mediating a chronic response during experimental arthritis. Am J Pathol 157:2081–2091

De Souza D, Fabri LJ, Nash A, Hilton DJ, Nicola NA, Baca M (2002) SH2 Domains from Suppressor of cytokine signaling-3 and protein tyrosine phosphatase SHP-2 have similar binding specificities. Biochemistry 41:9229–9236

DeChiara TM, Vejsada R, Poueymirou WT, Acheson A, Suri C, Conover JC, Friedman B, McClain J, Pan L, Stahl N, et al. (1995) Mice lacking the CNTF receptor, unlike mice lacking CNTF, exhibit profound motor neuron deficits at birth. Cell 83:313–322

Deller MC, Hudson KR, Ikemizu S, Bravo J, Jones EY, Heath JK (2000) Crystal structure and functional dissection of the cytostatic cytokine oncostatin M. Structure Fold Des 8:863–874

Deng C, Goluszko E, Tuzun E, Yang H, Christadoss P (2002) Resistance to experimental autoimmune myasthenia gravis in IL-6-deficient mice is associated with reduced germinal center formation and C3 production. J Immunol 169:1077–1083

DiCosmo BF, Geba GP, Picarella D, Elias JA, Rankin JA, Stripp BR, Whitsett JA, Flavell RA (1994a) Airway epithelial cell expression of interleukin-6 in transgenic mice. Uncoupling of airway inflammation and bronchial hyperreactivity. J Clin Invest 94:2028–2035

DiCosmo BF, Picarella D, Flavell RA (1994b) Local production of human IL-6 promotes insulitis but retards the onset of insulin-dependent diabetes mellitus in nonobese diabetic mice. Int Immunol 6:1829–1837

Diehl S, Anguita J, Hoffmeyer A, Zapton T, Ihle JN, Fikrig E, Rincon M (2000) Inhibition of Th1 differentiation by IL-6 is mediated by SOCS1. Immunity 13:805–815

Diehl S, Chow CW, Weiss L, Palmetshofer A, Twardzik T, Rounds L, Serfling E, Davis RJ, Anguita J, Rincon M (2002) Induction of NFATc2 expression by interleukin 6 promotes T helper type 2 differentiation. J Exp Med 196:39–49

Elson GC, Graber P, Losberger C, Herren S, Gretener D, Menoud LN, Wells TN, Kosco-Vilbois MH, Gauchat JF (1998) Cytokine-like factor-1, a novel soluble protein, shares homology with members of the cytokine type I receptor family. J Immunol 161:1371–1379

Elson GC, Lelievre E, Guillet C, Chevalier S, Plun-Favreau H, Froger J, Suard I, de Coignac AB, Delneste Y, Bonnefoy JY, Gauchat JF, Gascan H (2000) CLF associates with CLC to form a functional heteromeric ligand for the CNTF receptor complex. Nat Neurosci 3:867–872

Endo TA, Masuhara M, Yokouchi M, Suzuki R, Sakamoto H, Mitsui K, Matsumoto A, Tanimura S, Ohtsubo M, Misawa H, Miyazaki T, Leonor N, Taniguchi T, Fujita T, Kanakura Y, Komiya S, Yoshimura A (1997) A new protein containing an SH2 domain that inhibits JAK kinases. Nature 387:921–924

Ernst M, Inglese M, Waring P, Campbell IK, Bao S, Clay FJ, Alexander WS, Wicks IP, Tarlinton DM, Novak U, Heath JK, Dunn AR (2001) Defective gp130-mediated signal transducer and activator of transcription (STAT) signaling results in degenerative joint disease, gastrointestinal ulceration, and failure of uterine implantation. J Exp Med 194:189–203

Ershler WB, Keller ET (2000) Age-associated increased interleukin-6 gene expression, late-life diseases, and frailty. Annu Rev Med 51:245–270

Escary JL, Perreau J, Dumenil D, Ezine S, Brulet P (1993) Leukaemia inhibitory factor is necessary for maintenance of haematopoietic stem cells and thymocyte stimulation. Nature 363:361–364

Eugster HP, Frei K, Kopf M, Lassmann H, Fontana A (1998) IL-6-deficient mice resist myelin oligodendrocyte glycoprotein-induced autoimmune encephalomyelitis. Eur J Immunol 28:2178–2187

Feldmann M, Brennan FM, Maini RN (1996) Rheumatoid arthritis. Cell 85:307–310

Feng GS, Hui CC, Pawson T (1993) SH2-containing phosphotyrosine phosphatase as a target of protein-tyrosine kinases. Science 259:1607–1611

Frassanito MA, Cusmai A, Iodice G, Dammacco F (2001) Autocrine interleukin-6 production and highly malignant multiple myeloma: relation with resistance to drug-induced apoptosis. Blood 97:483–489

Fukada T, Hibi M, Yamanaka Y, Takahashi-Tezuka M, Fujitani Y, Yamaguchi T, Nakajima K, Hirano T (1996) Two signals are necessary for cell proliferation induced by a cytokine receptor gp130: involvement of STAT3 in antiapoptosis. Immunity 5:449–460

Fukada T, Ohtani T, Yoshida Y, Shirogane T, Nishida K, Nakajima K, Hibi M, Hirano T (1998) STAT3 orchestrates contradictory signals in cytokine-induced G1 to S cell-cycle transition. EMBO J 17:6670–6677

Fuller GM, Bunzel RJ, Woloski BM, Nham SU (1987) Isolation of hepatocyte stimulating factor from human monocytes. Biochem Biophys Res Commun 144:1003–1009

Gauldie J, Richards C, Harnish D, Lansdorp P, Baumann H (1987) Interferon $\beta2$/B-cell stimulatory factor type 2 shares identity with monocyte-derived hepatocyte-stimulating factor and regulates the major acute phase protein response in liver cells. Proc Natl Acad Sci USA 84:7251–7255

Gerhartz C, Heesel B, Sasse J, Hemmann U, Landgraf C, Schneider-Mergener J, Horn F, Heinrich PC, Graeve L (1996) Differential activation of acute phase response factor/STAT3 and STAT1 via the cytoplasmic domain of the interleukin 6 signal transducer gp130. I. Definition of a novel phosphotyrosine motif mediating STAT1 activation. J Biol Chem 271:12991–12998

Grisius MM, Bermudez DK, Fox PC (1997) Salivary and serum interleukin 6 in primary Sjogren's syndrome. J Rheumatol 24:1089–1091

Grotzinger J, Kurapkat G, Wollmer A, Kalai M, Rose-John S (1997) The family of the IL-6-type cytokines: specificity and promiscuity of the receptor complexes. Proteins 27:96–109

Haegeman G, Content J, Volckaert G, Derynck R, Tavernier J, Fiers W (1986) Structural analysis of the sequence coding for an inducible 26-kDa protein in human fibroblasts. Eur J Biochem 159:625–632

Heinrich PC, Castell JV, Andus T (1990) Interleukin-6 and the acute phase response. Biochem J 265:621–636

Hibi M, Murakami M, Saito M, Hirano T, Taga T, Kishimoto T (1990) Molecular cloning and expression of an IL-6 signal transducer, gp130. Cell 63:1149–1157

Hibi M, Hirano T (2000) Gab-family adapter molecules in signal transduction of cytokine and growth factor receptors, and T and B cell antigen receptors. Leuk Lymphoma 37:299–307

Hibi M, Hirano T (2001) IL-6 Receptor. In: Oppenheim JJ, Feldman M, Durum SK, Hirano T, Vilcek J, Nicola NA (eds) Cytokine reference. Academic Press, pp 1761–1778 vol 2)

Hirano T, Teranishi T, Toba H, Sakaguchi N, Fukukawa T, Tsuyuguchi I (1981) Human helper T cell factor(s) (ThF). I. Partial purification and characterization. J Immunol 126:517–522

Hirano T, Teranishi T, Lin B, Onoue K (1984a) Human helper T cell factor(s). IV. Demonstration of a human late-acting B cell differentiation factor acting on *Staphylococcus aureus* Cowan I- stimulated B cells. J Immunol 133:798–802

Hirano T, Teranishi T, Onoue K (1984b) Human helper T cell factor(s). III. Characterization of B cell differentiation factor I (BCDF I). J Immunol 132:229–234

Hirano T, Taga T, Nakano N, Yasukawa K, Kashiwamura S, Shimizu K, Nakajima K, Pyun KH, Kishimoto T (1985) Purification to homogeneity and characterization of human B-cell differentiation factor (BCDF or BSFp-2). Proc Natl Acad Sci USA 82:5490–5494

Hirano T, Yasukawa K, Harada H, Taga T, Watanabe Y, Matsuda T, Kashiwamura S, Nakajima K, Koyama K, Iwamatsu A, et al. (1986) Complementary DNA for a novel human interleukin (BSF-2) that induces B lymphocytes to produce immunoglobulin. Nature 324:73–76

Hirano T, Taga T, Yasukawa K, Nakajima K, Nakano N, Takatsuki F, Shimizu M, Murashima A, Tsunasawa S, Sakiyama F, et al. (1987) Human B-cell differentiation factor defined by an antipeptide antibody and its possible role in autoantibody production. Proc Natl Acad Sci USA 84:228–231

Hirano T, Matsuda T, Turner M, Miyasaka N, Buchan G, Tang B, Sato K, Shimizu M, Maini R, Feldmann M, et al. (1988) Excessive production of interleukin 6/B cell stimulatory factor-2 in rheumatoid arthritis. Eur J Immunol 18:1797–1801

Hirano T, Taga T, Yamasaki K, Matsuda T, Yasukawa K, Hirata Y, Yawata H, Tanabe O, Akira S, Kishimoto T (1989) Molecular cloning of the cDNAs for interleukin-6/B cell stimulatory factor 2 and its receptor. Ann N Y Acad Sci 557:167–178

Hirano T, Kishimoto T (1990) Interleukin-6. In: Sporn MB, Roberts AB (eds) Handbook of experimental pharmacology. (Peptide growth factors and their receptors vol 95) Springer-Verlag, pp 633–665

Hirano T (1991) Interleukin 6 (IL-6) and its receptor: their role in plasma cell neoplasias. Int J Cell Cloning 9:166–184

Hirano T, Nakajima K, Hibi M (1997) Signaling mechanisms through gp130: a model of the cytokine system. Cytokine Growth Factor Rev 8:241–252

Hirano T (1998) Interleukin 6 and its receptor: ten years later. Int Rev Immunol 16:249–284

Hirano T (1999) Molecular basis underlying functional pleiotropy of cytokines and growth factors. Biochem Biophys Res Commun 260:303–308

Hirano T, Ishihara K, Hibi M (2000) Roles of STAT3 in mediating the cell growth, differentiation and survival signals relayed through the IL-6 family of cytokine receptors. Oncogene 19:2548–2556

Hirano T, Fukada T (2001) IL-6 ligand and receptor family. In: Oppenheim JJ, Feldman M, Durum SK, Hirano T, Vilcek J, Nicola NA (eds) Cytokine reference. (vol 1) Academic Press, pp 523–535

Hirota H, Chen J, Betz UA, Rajewsky K, Gu Y, Ross J, Jr., Muller W, Chien KR (1999) Loss of a gp130 cardiac muscle cell survival pathway is a critical event in the onset of heart failure during biomechanical stress. Cell 97:189–198

Hof P, Pluskey S, Dhe-Paganon S, Eck MJ, Shoelson SE (1998) Crystal structure of the tyrosine phosphatase SHP-2. Cell 92:441–450

Holgado-Madruga M, Emlet DR, Moscatello DK, Godwin AK, Wong AJ (1996) A Grb2-associated docking protein in EGF- and insulin-receptor signaling. Nature 379:560–564

Holtkamp W, Stollberg T, Reis HE (1995) Serum interleukin-6 is related to disease activity but not disease specificity in inflammatory bowel disease. J Clin Gastroenterol 20:123–126

Ichiba M, Nakajima K, Yamanaka Y, Kiuchi N, Hirano T (1998) Autoregulation of the Stat3 gene through cooperation with a cAMP- responsive element-binding protein. J Biol Chem 273:6132–6138

Ichihara M, Hara T, Kim H, Murate T, Miyajima A (1997) Oncostatin M and leukemia inhibitory factor do not use the same functional receptor in mice. Blood 90:165–173

Ihara S, Nakajima K, Fukada T, Hibi M, Nagata S, Hirano T, Fukui Y (1997) Dual control of neurite outgrowth by STAT3 and MAP kinase in PC12 cells stimulated with interleukin-6. EMBO J 16:5345–5352

Itoh M, Yoshida Y, Nishida K, Narimatsu M, Hibi M, Hirano T (2000) Role of Gab1 in heart, placenta, and skin development and growth factor- and cytokine-induced extracellular signal-regulated kinase mitogen- activated protein kinase activation. Mol Cell Biol 20:3695–3704

Jilka RL, Hangoc G, Girasole G, Passeri G, Williams DC, Abrams JS, Boyce B, Broxmeyer H, Manolagas SC (1992) Increased osteoclast development after estrogen loss: mediation by interleukin-6. Science 257:88–91

Jones SA, Horiuchi S, Topley N, Yamamoto N, Fuller GM (2001) The soluble interleukin 6 receptor: mechanisms of production and implications in disease. FASEB J 15:43–58

Jostock T, Mullberg J, Ozbek S, Atreya R, Blinn G, Voltz N, Fischer M, Neurath MF, Rose-John S (2001) Soluble gp130 is the natural inhibitor of soluble interleukin-6 receptor transsignaling responses. Eur J Biochem 268:160–167

Kamimura D, Fu D, Matsuda Y, Atsumi T, Ohtani T, Park SJ, Ishihara K, Hirano T (2002) Tyrosine 759 of the cytokine receptor gp130 is involved in Listeria monocytogenes susceptibility. Genes Immun 3:136–143

Karaghiosoff M, Neubauer H, Lassnig C, Kovarik P, Schindler H, Pircher H, McCoy B, Bogdan C, Decker T, Brem G, Pfeffer K, Muller M (2000) Partial impairment of cytokine responses in Tyk2-deficient mice. Immunity 13:549–560

Kawano M, Hirano T, Matsuda T, Taga T, Horii Y, Iwato K, Asaoku H, Tang B, Tanabe O, Tanaka H, et al. (1988) Autocrine generation and requirement of BSF-2/IL-6 for human multiple myelomas. Nature 332:83–85

Kawasaki K, Gao YH, Yokose S, Kaji Y, Nakamura T, Suda T, Yoshida K, Taga T, Kishimoto T, Kataoka H, Yuasa T, Norimatsu H, Yamaguchi A (1997) Osteoclasts are present in gp130-deficient mice. Endocrinology 138:4959–4965

Kim H, Hawley TS, Hawley RG, Baumann H (1998) Protein tyrosine phosphatase 2 (SHP-2) moderates signaling by gp130 but is not required for the induction of acute-phase plasma protein genes in hepatic cells. Mol Cell Biol 18:1525–1533

Kim H, Baumann H (1999) Dual signaling role of the protein tyrosine phosphatase SHP-2 in regulating expression of acute-phase plasma proteins by interleukin-6 cytokine receptors in hepatic cells. Mol Cell Biol 19:5326–5338

Klein B, Zhang XG, Lu ZY, Bataille R (1995) Interleukin-6 in human multiple myeloma. Blood 85:863–872

Kopf M, Baumann H, Freer G, Freudenberg M, Lamers M, Kishimoto T, Zinkernagel R, Bluethmann H, Kohler G (1994) Impaired immune and acute-phase responses in interleukin-6-deficient mice. Nature 368:339–342

Kumanogoh A, Marukawa S, Kumanogoh T, Hirota H, Yoshida K, Lee IS, Yasui T, Taga T, Kishimoto T (1997) Impairment of antigen-specific antibody production in transgenic mice expressing a dominant-negative form of gp130. Proc Natl Acad Sci USA 94:2478–2482

La Flamme AC, Pearce EJ (1999) The absence of IL-6 does not affect Th2 cell development in vivo, but does lead to impaired proliferation, IL-2 receptor expression, and B cell responses. J Immunol 162:5829–5837

Le JM, Vilcek J (1989) Interleukin 6: a multifunctional cytokine regulating immune reactions and the acute phase protein response. Lab Invest 61:588–602

Lefebvre O, Chenard MP, Masson R, Linares J, Dierich A, LeMeur M, Wendling C, Tomasetto C, Chambon P, Rio MC (1996) Gastric mucosa abnormalities and tumorigenesis in mice lacking the pS2 trefoil protein. Science 274:259–262

Lehmann U, Schmitz J, Weissenbach M, Sobota RM, Hortner M, Friederichs K, Behrmann I, Tsiaris W, Sasaki A, Schneider-Mergener J, Yoshimura A, Neel BG, Heinrich PC, Schaper F (2002) SHP2 and SOCS3 contribute to Y759-dependent attenuation of IL-6- signaling through gp130. J Biol Chem 27:27

Lelievre E, Plun-Favreau H, Chevalier S, Froger J, Guillet C, Elson GC, Gauchat JF, Gascan H (2001) Signaling pathways recruited by the cardiotrophin-like cytokine/cytokine-like factor-1 composite cytokine: specific requirement of the membrane-bound form of ciliary neurotrophic factor receptor αcomponent. J Biol Chem 276:22476–22484

Linker RA, Maurer M, Gaupp S, Martini R, Holtmann B, Giess R, Rieckmann P, Lassmann H, Toyka KV, Sendtner M, Gold R (2002) CNTF is a major protective factor in demyelinating CNS disease: A neurotrophic cytokine as modulator in neuroinflammation. Nat Med 8:620–624

Liu B, Liao J, Rao X, Kushner SA, Chung CD, Chang DD, Shuai K (1998) Inhibition of Stat1-mediated gene activation by PIAS1. Proc Natl Acad Sci USA 95:10626–10631

Liu F, Poursine-Laurent J, Wu HY, Link DC (1997) Interleukin-6 and the granulocyte colony-stimulating factor receptor are major independent regulators of granulopoiesis in vivo but are not required for lineage commitment or terminal differentiation. Blood 90:2583–2590

Liu Y, Rohrschneider LR (2002) The gift of Gab. FEBS Lett 515:1–7

Liu Z, Simpson RJ, Cheers C (1995) Interaction of interleukin-6, tumor necrosis factor and interleukin-1 during Listeria infection. Immunology 85:562–567

Lutticken C, Wegenka UM, Yuan J, Buschmann J, Schindler C, Ziemiecki A, Harpur AG, Wilks AF, Yasukawa K, Taga T, et al. (1994) Association of transcription factor APRF and protein kinase Jak1 with the interleukin-6 signal transducer gp130. Science 263:89–92

Maddox L, Schwartz DA (2002) The pathophysiology of asthma. Annu Rev Med 53:477–498

Martens AS, Bode JG, Heinrich PC, Graeve L (2000) The cytoplasmic domain of the interleukin-6 receptor gp80 mediates its basolateral sorting in polarized madin-darby canine kidney cells. J Cell Sci 113:3593–3602

Masu Y, Wolf E, Holtmann B, Sendtner M, Brem G, Thoenen H (1993) Disruption of the CNTF gene results in motor neuron degeneration. Nature 365:27–32

May LT, Helfgott DC, Sehgal PB (1986) Anti-β-interferon antibodies inhibit the increased expression of HLA- B7 mRNA in tumor necrosis factor-treated human fibroblasts: structural studies of the β2 interferon involved. Proc Natl Acad Sci USA 83:8957–8961

May LT, Ghrayeb J, Santhanam U, Tatter SB, Sthoeger Z, Helfgott DC, Chiorazzi N, Grieninger G, Sehgal PB (1988a) Synthesis and secretion of multiple forms of β2-interferon/B-cell differentiation factor 2/hepatocyte-stimulating factor by human fibroblasts and monocytes. J Biol Chem 263:7760–7766

May LT, Santhanam U, Tatter SB, Bhardwaj N, Ghrayeb J, Sehgal PB (1988b) Phosphorylation of secreted forms of human β2-interferon/hepatocyte stimulating factor/interleukin-6. Biochem Biophys Res Commun 152:1144–1150

Mihara M, Moriya Y, Kishimoto T, Ohsugi Y (1995) Interleukin-6 (IL-6) induces the proliferation of synovial fibroblastic cells in the presence of soluble IL-6 receptor. Br J Rheumatol 34:321–325

Miles SA, Rezai AR, Salazar-Gonzalez JF, Vander Meyden M, Stevens RH, Logan DM, Mitsuyasu RT, Taga T, Hirano T, Kishimoto T, et al. (1990) AIDS Kaposi sarcoma-derived cells produce and respond to interleukin 6. Proc Natl Acad Sci USA 87:4068–4072

Mitani H, Katayama N, Araki H, Ohishi K, Kobayashi K, Suzuki H, Nishii K, Masuya M, Yasukawa K, Minami N, Shiku H (2000) Activity of interleukin 6 in the differentiation of monocytes to macrophages and dendritic cells. Br J Haematol 109:288–295

Molnar EL, Hegyesi H, Toth S, Darvas Z, Laszlo V, Szalai C, Falus A (2000) Biosynthesis of interleukin-6, an autocrine growth factor for melanoma, is regulated by melanoma-derived histamine. Semin Cancer Biol 10:25–28

Moritz RL, Ward LD, Tu GF, Fabri LJ, Ji H, Yasukawa K, Simpson RJ (1999) The N-terminus of gp130 is critical for the formation of the high- affinity interleukin-6 receptor complex. Growth Factors 16:265–278

Moshage H (1997) Cytokines and the hepatic acute phase response. J Pathol 181:257–266

Muraguchi A, Hirano T, Tang B, Matsuda T, Horii Y, Nakajima K, Kishimoto T (1988) The essential role of B cell stimulatory factor 2 (BSF-2/IL-6) for the terminal differentiation of B cells. J Exp Med 167:332–344

Murakami M, Hibi M, Nakagawa N, Nakagawa T, Yasukawa K, Yamanishi K, Taga T, Kishimoto T (1993) IL-6-induced homodimerization of gp130 and associated activation of a tyrosine kinase. Science 260:1808–1810

Naka T, Narazaki M, Hirata M, Matsumoto T, Minamoto S, Aono A, Nishimoto N, Kajita T, Taga T, Yoshizaki K, Akira S, Kishimoto T (1997) Structure and function of a new STAT-induced STAT inhibitor. Nature 387:924–929

Nakajima K, Yamanaka Y, Nakae K, Kojima H, Ichiba M, Kiuchi N, Kitaoka T, Fukada T, Hibi M, Hirano T (1996) A central role for Stat3 in IL-6-induced regulation of growth and differentiation in M1 leukemia cells. EMBO J 15:3651–3658

Nandurkar HH, Robb L, Tarlinton D, Barnett L, Kontgen F, Begley CG (1997) Adult mice with targeted mutation of the interleukin-11 receptor (IL11Ra) display normal hematopoiesis. Blood 90:2148–2159

Narazaki M, Witthuhn BA, Yoshida K, Silvennoinen O, Yasukawa K, Ihle JN, Kishimoto T, Taga T (1994) Activation of JAK2 kinase mediated by the interleukin 6 signal transducer gp130. Proc Natl Acad Sci USA 91:2285–2289

Narimatsu M, Maeda H, Itoh S, Atsumi T, Ohtani T, Nishida K, Itoh M, Kamimura D, Park SJ, Mizuno K, Miyazaki J, Hibi M, Ishihara K, Nakajima K, Hirano T (2001) Tissue-specific autoregulation of the stat3 gene and its role in interleukin-6-induced survival signals in T cells. Mol Cell Biol 21:6615–6625

Nicholas J, Ruvolo VR, Burns WH, Sandford G, Wan X, Ciufo D, Hendrickson SB, Guo HG, Hayward GS, Reitz MS (1997) Kaposi's sarcoma-associated human herpesvirus-8 encodes homologs of macrophage inflammatory protein-1 and interleukin-6. Nat Med 3:287–292

Nicholson SE, De Souza D, Fabri LJ, Corbin J, Willson TA, Zhang JG, Silva A, Asimakis M, Farley A, Nash AD, Metcalf D, Hilton DJ, Nicola NA, Baca M (2000) Suppressor of cytokine signaling-3 preferentially binds to the SHP-2- binding site on the shared cytokine receptor subunit gp130. Proc Natl Acad Sci USA 97:6493–6498

Nishida K, Yoshida Y, Itoh M, Fukada T, Ohtani T, Shirogane T, Atsumi T, Takahashi-Tezuka M, Ishihara K, Hibi M, Hirano T (1999) Gab-family adapter proteins act downstream of cytokine and growth factor receptors and T- and B-cell antigen receptors. Blood 93:1809–1816

Noda M, Takeda K, Sugimoto H, Hosoi T, Takechi K, Hara T, Ishikawa H, Arimura H, Konno K (1991) Purification and characterization of human fibroblast derived differentiation inducing factor for human monoblastic leukemia cells identical to interleukin-6. Anticancer Res 11:961–968

O'Barr S, Cooper NR (2000) The C5a complement activation peptide increases IL-1β and IL-6 release from amyloid-β-primed human monocytes: implications for Alzheimer's disease. J Neuroimmunol 109:87–94

O'Brien CA, Manolagas SC (1997) Isolation and characterization of the human gp130 promoter. Regulation by STATS. J Biol Chem 272:15003–15010

Ohshima S, Saeki Y, Mima T, Sasai M, Nishioka K, Nomura S, Kopf M, Katada Y, Tanaka T, Suemura M, Kishimoto T (1998) Interleukin 6 plays a key role in the development of antigen-induced arthritis. Proc Natl Acad Sci USA 95:8222–8226

Ohtani T, Ishihara K, Atsumi T, Nishida K, Kaneko Y, Miyata T, Itoh S, Narimatsu M, Maeda H, Fukada T, Itoh M, Okano H, Hibi M, Hirano T (2000) Dissection of signaling cascades through gp130 in vivo: reciprocal roles for STAT3- and SHP2-mediated signals in immune responses. Immunity 12:95–105

Okuda Y, Sakoda S, Bernard CC, Fujimura H, Saeki Y, Kishimoto T, Yanagihara T (1998) IL-6-deficient mice are resistant to the induction of experimental autoimmune encephalomyelitis provoked by myelin oligodendrocyte glycoprotein. Int Immunol 10:703–708

Oppenheim RW, Wiese S, Prevette D, Armanini M, Wang S, Houenou LJ, Holtmann B, Gotz R, Pennica D, Sendtner M (2001) Cardiotrophin-1, a muscle-derived cytokine, is required for the survival of subpopulations of developing motoneurons. J Neurosci 21:1283–1291

O'Shea JJ, Gadina M, Schreiber RD (2002) Cytokine signaling in 2002: new surprises in the Jak/Stat pathway. Cell 109 Suppl:S121–131

Papassotiropoulos A, Hock C, Nitsch RM (2001) Genetics of interleukin 6: implications for Alzheimer's disease. Neurobiol Aging 22:863–871

Parganas E, Wang D, Stravopodis D, Topham DJ, Marine JC, Teglund S, Vanin EF, Bodner S, Colamonici OR, van Deursen JM, Grosveld G, Ihle JN (1998) Jak2 is essential for signaling through a variety of cytokine receptors. Cell 93:385–395

Pedersen BK, Steensberg A, Schjerling P (2001) Muscle-derived interleukin-6: possible biological effects. J Physiol 536:329–337

Peters M, Schirmacher P, Goldschmitt J, Odenthal M, Peschel C, Fattori E, Ciliberto G, Dienes HP, Meyer zum Buschenfelde KH, Rose-John S (1997) Extramedullary expansion of hematopoietic progenitor cells in interleukin (IL)-6-sIL-6R double transgenic mice. J Exp Med 185:755–766

Peters M, Muller AM, Rose-John S (1998) Interleukin-6 and soluble interleukin-6 receptor: direct stimulation of gp130 and hematopoiesis. Blood 92:3495–3504

Poli V, Balena R, Fattori E, Markatos A, Yamamoto M, Tanaka H, Ciliberto G, Rodan GA, Costantini F (1994) Interleukin-6 deficient mice are protected from bone loss caused by estrogen depletion. EMBO J 13:1189–1196

Pradhan AD, Manson JE, Rifai N, Buring JE, Ridker PM (2001) C-reactive protein, interleukin 6, and risk of developing type 2 diabetes mellitus. JAMA 286:327–334

Ramsay AJ, Husband AJ, Ramshaw IA, Bao S, Matthaei KI, Koehler G, Kopf M (1994) The role of interleukin-6 in mucosal IgA antibody responses in vivo. Science 264:561–563

Rifas L (1999) Bone and cytokines: beyond IL-1, IL-6 and TNF-α. Calcif Tissue Int 64:1–7

Rincon M, Anguita J, Nakamura T, Fikrig E, Flavell RA (1997) Interleukin (IL)-6 directs the differentiation of IL-4-producing CD4+ T cells. J Exp Med 185:461–469

Robb L, Li R, Hartley L, Nandurkar HH, Koentgen F, Begley CG (1998) Infertility in female mice lacking the receptor for interleukin 11 is due to a defective uterine response to implantation. Nat Med 4:303–308

Robledo O, Fourcin M, Chevalier S, Guillet C, Auguste P, Pouplard-Barthelaix A, Pennica D, Gascan H (1997) Signaling of the cardiotrophin-1 receptor. Evidence for a third receptor component. J Biol Chem 272:4855–4863

Rodig SJ, Meraz MA, White JM, Lampe PA, Riley JK, Arthur CD, King KL, Sheehan KC, Yin L, Pennica D, Johnson EM Jr, Schreiber RD (1998) Disruption of the Jak1 gene demonstrates obligatory and nonredundant roles of the Jaks in cytokine-induced biologic responses. Cell 93:373–383

Romani L, Mencacci A, Cenci E, Spaccapelo R, Toniatti C, Puccetti P, Bistoni F, Poli V (1996) Impaired neutrophil response and CD4+ T helper cell 1 development in interleukin 6-deficient mice infected with Candida albicans. J Exp Med 183:1345–1355

Romano M, Sironi M, Toniatti C, Polentarutti N, Fruscella P, Ghezzi P, Faggioni R, Luini W, van Hinsbergh V, Sozzani S, Bussolino F, Poli V, Ciliberto G, Mantovani A (1997) Role of IL-6 and its soluble receptor in induction of chemokines and leukocyte recruitment. Immunity 6:315–325

Saito M, Yoshida K, Hibi M, Taga T, Kishimoto T (1992) Molecular cloning of a murine IL-6 receptor-associated signal transducer, gp130, and its regulated expression in vivo. J Immunol 148:4066–4071

Samoilova EB, Horton JL, Hilliard B, Liu TS, Chen Y (1998) IL-6-deficient mice are resistant to experimental autoimmune encephalomyelitis: roles of IL-6 in the activation and differentiation of autoreactive T cells. J Immunol 161:6480–6486

Santhanam U, Ray A, Sehgal PB (1991) Repression of the interleukin 6 gene promoter by p53 and the retinoblastoma susceptibility gene product. Proc Natl Acad Sci USA 88:7605–7609

Sasai M, Saeki Y, Ohshima S, Nishioka K, Mima T, Tanaka T, Katada Y, Yoshizaki K, Suemura M, Kishimoto T (1999) Delayed onset and reduced severity of collagen-induced arthritis in interleukin-6-deficient mice. Arthritis Rheum 42:1635–1643

Schaper F, Gendo C, Eck M, Schmitz J, Grimm C, Anhuf D, Kerr IM, Heinrich PC (1998) Activation of the protein tyrosine phosphatase SHP2 via the interleukin- 6 signal transducing receptor protein gp130 requires tyrosine kinase Jak1 and limits acute-phase protein expression. Biochem J 335:557–565

Schirmacher P, Peters M, Ciliberto G, Blessing M, Lotz J, Meyer zum Buschenfelde KH, Rose-John S (1998) Hepatocellular hyperplasia, plasmacytoma formation, and extramedullary hematopoiesis in interleukin (IL)-6/soluble IL-6 receptor double-transgenic mice. Am J Pathol 153:639–648

Schmitz J, Dahmen H, Grimm C, Gendo C, Muller-Newen G, Heinrich PC, Schaper F (2000a) The cytoplasmic tyrosine motifs in full-length glycoprotein 130 have different roles in IL-6 signal transduction. J Immunol 164:848–854

Schmitz J, Weissenbach M, Haan S, Heinrich PC, Schaper F (2000b) SOCS3 exerts its inhibitory function on interleukin-6 signal transduction through the SHP2 recruitment site of gp130. J Biol Chem 275:12848–12856

Schreiber M, Kolbus A, Piu F, Szabowski A, Mohle-Steinlein U, Tian J, Karin M, Angel P, Wagner EF (1999) Control of cell cycle progression by c-Jun is p53 dependent. Genes Dev 13:607–619

Sehgal PB, Grienger G, Tosato G (1989) Regulation of the acute phase and immune responses: interleukin-6. Ann N Y Acad Sci 557:1–583

Senaldi G, Stolina M, Guo J, Faggioni R, McCabe S, Kaufman SA, Van G, Xu W, Fletcher FA, Boone T, Chang MS, Sarmiento U, Cattley RC (2002) Regulatory effects of novel neurotrophin-1/b cell-stimulating factor-3 (cardiotrophin-like cytokine) on B cell function. J Immunol 168:5690–5698

Shen MM, Skoda RC, Cardiff RD, Campos-Torres J, Leder P, Ornitz DM (1994) Expression of LIF in transgenic mice results in altered thymic epithelium and apparent interconversion of thymic and lymph node morphologies. EMBO J 13:1375–1385

Shimoda K, Kato K, Aoki K, Matsuda T, Miyamoto A, Shibamori M, Yamashita M, Numata A, Takase K, Kobayashi S, Shibata S, Asano Y, Gondo H, Sekiguchi K, Nakayama K, Nakayama T, Okamura T, Okamura S, Niho Y (2000) Tyk2 plays a restricted role in IFN α signaling, although it is required for IL-12-mediated T cell function. Immunity 13:561–571

Shirogane T, Fukada T, Muller JM, Shima DT, Hibi M, Hirano T (1999) Synergistic roles for Pim-1 and c-Myc in STAT3-mediated cell cycle progression and antiapoptosis. Immunity 11:709–719

Shirota K, LeDuy L, Yuan SY, Jothy S (1990) Interleukin-6 and its receptor are expressed in human intestinal epithelial cells. Virchows Arch B Cell Pathol Incl Mol Pathol 58:303–308

Smith PC, Hobisch A, Lin DL, Culig Z, Keller ET (2001) Interleukin-6 and prostate cancer progression. Cytokine Growth Factor Rev 12:33–40

Somers W, Stahl M, Seehra JS (1997) 1.9 A crystal structure of interleukin 6: implications for a novel mode of receptor dimerization and signaling. EMBO J 16:989–997

Songyang Z, Shoelson SE, Chaudhuri M, Gish G, Pawson T, Haser WG, King F, Roberts T, Ratnofsky S, Lechleider RJ, et al. (1993) SH2 domains recognize specific phosphopeptide sequences. Cell 72:767–778

Stahl N, Boulton TG, Farruggella T, Ip NY, Davis S, Witthuhn BA, Quelle FW, Silvennoinen O, Barbieri G, Pellegrini S, et al. (1994) Association and activation of Jak-Tyk kinases by CNTF-LIF-OSM-IL-6 β receptor components. Science 263:92–95

Stahl N, Farruggella TJ, Boulton TG, Zhong Z, Darnell JE, Jr., Yancopoulos GD (1995) Choice of STATs and other substrates specified by modular tyrosine- based motifs in cytokine receptors. Science 267:1349-1353

Starr R, Willson TA, Viney EM, Murray LJ, Rayner JR, Jenkins BJ, Gonda TJ, Alexander WS, Metcalf D, Nicola NA, Hilton DJ (1997) A family of cytokine-inducible inhibitors of signaling. Nature 387:917–921

Stelmasiak Z, Koziol-Montewka M, Dobosz B, Rejdak K, Bartosik-Psujek H, Mitosek-Szewczyk K, Belniak-Legiec E (2000) Interleukin-6 concentration in serum and cerebrospinal fluid in multiple sclerosis patients. Med Sci Monit 6:1104–1108

Stewart CL, Kaspar P, Brunet LJ, Bhatt H, Gadi I, Kontgen F, Abbondanzo SJ (1992) Blastocyst implantation depends on maternal expression of leukaemia inhibitory factor. Nature 359:76–79

Stuart RA, Littlewood AJ, Maddison PJ, Hall ND (1995) Elevated serum interleukin-6 levels associated with active disease in systemic connective tissue disorders. Clin Exp Rheumatol 13:17–22

Suematsu S, Matsuda T, Aozasa K, Akira S, Nakano N, Ohno S, Miyazaki J, Yamamura K, Hirano T, Kishimoto T (1989) IgG1 plasmacytosis in interleukin 6 transgenic mice. Proc Natl Acad Sci USA 86:7547-7551

Suematsu S, Matsusaka T, Matsuda T, Ohno S, Miyazaki J, Yamamura K, Hirano T, Kishimoto T (1992) Generation of plasmacytomas with the chromosomal translocation t(12;15) in interleukin 6 transgenic mice. Proc Natl Acad Sci USA 89:232–235

Suzuki A, Hanada T, Mitsuyama K, Yoshida T, Kamizono S, Hoshino T, Kubo M, Yamashita A, Okabe M, Takeda K, Akira S, Matsumoto S, Toyonaga A, Sata M, Yoshimura A (2001) CIS3/SOCS3/SSI3 plays a negative regulatory role in STAT3 activation and intestinal inflammation. J Exp Med 193:471–481

Taga T, Hibi M, Hirata Y, Yamasaki K, Yasukawa K, Matsuda T, Hirano T, Kishimoto T (1989) Interleukin-6 triggers the association of its receptor with a possible signal transducer, gp130. Cell 58:573–581

Takahashi-Tezuka M, Yoshida Y, Fukada T, Ohtani T, Yamanaka Y, Nishida K, Nakajima K, Hibi M, Hirano T (1998) Gab1 acts as an adapter molecule linking the cytokine receptor gp130 to ERK mitogen-activated protein kinase. Mol Cell Biol 18:4109–4117

Takeda K, Kaisho T, Yoshida N, Takeda J, Kishimoto T, Akira S (1998) Stat3 activation is responsible for IL-6-dependent T cell proliferation through preventing apoptosis: generation and characterization of T cell- specific Stat3-deficient mice. J Immunol 161:4652–4660

Tanaka M, Kishimura M, Ozaki S, Osakada F, Hashimoto H, Okubo M, Murakami M, Nakao K (2000) Cloning of novel soluble gp130 and detection of its neutralizing autoantibodies in rheumatoid arthritis. J Clin Invest 106:137–144

Tanaka T, Katada Y, Higa S, Fujiwara H, Wang W, Saeki Y, Ohshima S, Okuda Y, Suemura M, Kishimoto T (2001) Enhancement of T helper2 response in the absence of interleukin (IL-)6; an inhibition of IL-4-mediated T helper2 cell differentiation by IL-6. Cytokine 13:193–201

Tebbutt NC, Giraud AS, Inglese M, Jenkins B, Waring P, Clay FJ, Malki S, Alderman BM, Grail D, Hollande F, Heath JK, Ernst M (2002) Reciprocal regulation of gastrointestinal homeostasis by SHP2 and STAT- mediated trefoil gene activation in gp130 mutant mice. Nat Med 8:1089–1097

ten Hoeve J, de Jesus Ibarra-Sanchez M, Fu Y, Zhu W, Tremblay M, David M, Shuai K (2002) Identification of a nuclear Stat1 protein tyrosine phosphatase. Mol Cell Biol 22:5662–5668

Teranishi T, Hirano T, Arima N, Onoue K (1982) Human helper T cell factor(s) (ThF). II. Induction of IgG production in B lymphoblastoid cell lines and identification of T cell-replacing factor- (TRF) like factor(s). J Immunol 128:1903–1908

Tonouchi N, Oouchi N, Kashima N, Kawai M, Nagase K, Okano A, Matsui H, Yamada K, Hirano T, Kishimoto T (1988) High-level expression of human BSF-2/IL-6 cDNA in *Escherichia coli* using a new type of expression-preparation system. J Biochem (Tokyo) 104:30–34

Tsunenari T, Koishihara Y, Nakamura A, Moriya M, Ohkawa H, Goto H, Shimazaki C, Nakagawa M, Ohsugi Y, Kishimoto T, Akamatsu K (1997) New xenograft model of multiple myeloma and efficacy of a humanized antibody against human interleukin-6 receptor. Blood 90:2437–2444

Udagawa N, Takahashi N, Katagiri T, Tamura T, Wada S, Findlay DM, Martin TJ, Hirota H, Taga T, Kishimoto T, et al. (1995) Interleukin (IL)-6 induction of osteoclast differentiation depends on IL-6 receptors expressed on osteoblastic cells but not on osteoclast progenitors. J Exp Med 182:1461–1468

Ueda T, Bruchovsky N, Sadar MD (2002) Activation of the androgen receptor N-terminal domain by interleukin-6 via MAPK and STAT3 signal transduction pathways. J Biol Chem 277:7076–7085

Uozumi H, Hiroi Y, Zou Y, Takimoto E, Toko H, Niu P, Shimoyama M, Yazaki Y, Nagai R, Komuro I (2001) gp130 plays a critical role in pressure overload-induced cardiac hypertrophy. J Biol Chem 276:23115–23119

Van Snick J, Cayphas S, Szikora JP, Renauld JC, Van Roost E, Boon T, Simpson RJ (1988) cDNA cloning of murine interleukin-HP1: homology with human interleukin 6. Eur J Immunol 18:193–197

Van Snick J (1990) Interleukin-6: an overview. Annu Rev Immunol 8:253–278

Vink A, Coulie PG, Wauters P, Nordan RP, Van Snick J (1988) B cell growth and differentiation activity of interleukin-HP1 and related murine plasmacytoma growth factors. Synergy with interleukin 1. Eur J Immunol 18:607–612

Vogel W, Lammers R, Huang J, Ullrich A (1993) Activation of a phosphotyrosine phosphatase by tyrosine phosphorylation. Science 259:1611–1614

Wallenius K, Wallenius V, Sunter D, Dickson SL, Jansson JO (2002a) Intracerebroventricular interleukin-6 treatment decreases body fat in rats. Biochem Biophys Res Commun 293:560–565

Wallenius V, Wallenius K, Ahren B, Rudling M, Carlsten H, Dickson SL, Ohlsson C, Jansson JO (2002b) Interleukin-6-deficient mice develop mature-onset obesity. Nat Med 8:75–79

Wang J, Homer RJ, Chen Q, Elias JA (2000a) Endogenous and exogenous IL-6 inhibit aeroallergen-induced Th2 inflammation. J Immunol 165:4051–4061

Wang J, Homer RJ, Hong L, Cohn L, Lee CG, Jung S, Elias JA (2000b) IL-11 selectively inhibits aeroallergen-induced pulmonary eosinophilia and Th2 cytokine production. J Immunol 165:2222–2231

Wang YD, Gu ZJ, Huang JA, Zhu YB, Zhou ZH, Xie W, Xu Y, Qiu YH, Zhang XG (2002) gp130-linked signal transduction promotes the differentiation and maturation of dendritic cells. Int Immunol 14:599–603

Ward LD, Howlett GJ, Discolo G, Yasukawa K, Hammacher A, Moritz RL, Simpson RJ (1994) High affinity interleukin-6 receptor is a hexameric complex consisting of two molecules each of interleukin-6, interleukin-6 receptor, and gp-130. J Biol Chem 269:23286–23289

Ware CB, Horowitz MC, Renshaw BR, Hunt JS, Liggitt D, Koblar SA, Gliniak BC, McKenna HJ, Papayannopoulou T, Thoma B, et al. (1995) Targeted disruption of the low-affinity leukemia inhibitory factor receptor gene causes placental, skeletal, neural and metabolic defects and results in perinatal death. Development 121:1283–1299

Weissenbach J, Chernajovsky Y, Zeevi M, Shulman L, Soreq H, Nir U, Wallach D, Perricaudet M, Tiollais P, Revel M (1980) Two interferon mRNAs in human fibroblasts: in vitro translation and *Escherichia coli* cloning studies. Proc Natl Acad Sci USA 77:7152–7156

Wolf I, Jenkins BJ, Liu Y, Seiffert M, Custodio JM, Young P, Rohrschneider LR (2002) Gab3, a new DOS/Gab family member, facilitates macrophage differentiation. Mol Cell Biol 22:231–244

Wong CK, Ho CY, Ko FW, Chan CH, Ho AS, Hui DS, Lam CW (2001) Proinflammatory cytokines (IL-17, IL-6, IL-18 and IL-12) and Th cytokines (IFN-γ, IL-4, IL-10 and IL-13) in patients with allergic asthma. Clin Exp Immunol 125:177–183

Xing Z, Gauldie J, Cox G, Baumann H, Jordana M, Lei X-F, Achong M (1998) IL-6 is an antiinflammatory cytokine required for controlling local or systemic acute inflammatory responses. J. Clin. Invest. 101:311–320

Yamanaka Y, Nakajima K, Fukada T, Hibi M, Hirano T (1996) Differentiation and growth arrest signals are generated through the cytoplasmic region of gp130 that is essential for Stat3 activation. EMBO J 15:1557–1565

Yamasaki K, Taga T, Hirata Y, Yawata H, Kawanishi Y, Seed B, Taniguchi T, Hirano T, Kishimoto T (1988) Cloning and expression of the human interleukin-6 (BSF-2/IFN β 2) receptor. Science 241:825–828

Yasukawa H, Sasaki A, Yoshimura A (2000) Negative regulation of cytokine signaling pathways. Annu Rev Immunol 18:143–164

Yasukawa K, Hirano T, Watanabe Y, Muratani K, Matsuda T, Nakai S, Kishimoto T (1987) Structure and expression of human B cell stimulatory factor-2 (BSF-2/IL- 6) gene. EMBO J 6:2939–2945

Yokoyama A, Kohno N, Fujino S, Hamada H, Inoue Y, Fujioka S, Ishida S, Hiwada K (1995) Circulating interleukin-6 levels in patients with bronchial asthma. Am J Respir Crit Care Med 151:1354–1358

Yoshida K, Taga T, Saito M, Suematsu S, Kumanogoh A, Tanaka T, Fujiwara H, Hirata M, Yamagami T, Nakahata T, Hirabayashi T, Yoneda Y, Tanaka K, Wang WZ, Mori C, Shiota K, Yoshida N, Kishimoto T (1996) Targeted disruption of gp130, a common signal transducer for the interleukin 6 family of cytokines, leads to myocardial and hematological disorders. Proc Natl Acad Sci USA 93:407–411

Yoshizaki K, Matsuda T, Nishimoto N, Kuritani T, Taeho L, Aozasa K, Nakahata T, Kawai H, Tagoh H, Komori T, et al. (1989) Pathogenic significance of interleukin-6 (IL-6/BSF-2) in Castleman's disease. Blood 74:1360–1367

Yudkin JS, Kumari M, Humphries SE, Mohamed-Ali V (2000) Inflammation, obesity, stress and coronary heart disease: is interleukin-6 the link? Atherosclerosis 148:209–214

Zhang JG, Zhang Y, Owczarek CM, Ward LD, Moritz RL, Simpson RJ, Yasukawa K, Nicola NA (1998) Identification and characterization of two distinct truncated forms of gp130 and a soluble form of leukemia inhibitory factor receptor α-chain in normal human urine and plasma. J Biol Chem 273:10798–10805

Zhong J, Dietzel ID, Wahle P, Kopf M, Heumann R (1999) Sensory impairments and delayed regeneration of sensory axons in interleukin-6-deficient mice. J Neurosci 19:4305–4313

Zhu M, Oishi K, Lee SC, Patterson PH (2001) Studies using leukemia inhibitory factor (LIF) knockout mice and a LIF adenoviral vector demonstrate a key anti-inflammatory role for this cytokine in cutaneous inflammation. J Immunol 166:2049–2054

Zilberstein A, Ruggieri R, Korn JH, Revel M (1986) Structure and expression of cDNA and genes for human interferon-β-2, a distinct species inducible by growth-stimulatory cytokines. EMBO J 5:2529–2537

M. Tanaka · A. Miyajima

Oncostatin M, a multifunctional cytokine

Published online: 17 June 2003
© Springer-Verlag 2003

Abstract Oncostatin M (OSM) is a multifunctional cytokine that belongs to the Interleukin (IL)-6 subfamily. Among the family members, OSM is most closely related to leukemia inhibitory factor (LIF) and it in fact utilizes the LIF receptor in addition to its specific receptor in the human. While OSM was originally recognized by its unique activity to inhibit the proliferation of tumor cells, accumulating evidence now indicates that OSM exhibits many unique biological activities in inflammation, hematopoiesis, and development. Here, we review the profile of OSM activities, receptors, and signal transduction.

Abbreviations *G-CSF:* Granulocyte-colony stimulating factor · *GM-CSF:* Granulocyte-macrophage colony stimulating factor · *GAS:* γ Interferon-activated site · *Grb:* Growth factor receptor-bound protein · *Gab:* Grb-2-associated binder · *SHP:* SH2 domain-containing protein tyrosine phosphatase · *STAT:* Signal transducer and activator of transcription · *TEL:* Translocated ets leukemia

Introduction

In the last two decades, a large number of cytokines have been found and are classified into several families based on their structural properties as well as their receptor components. Oncostatin M (OSM) is a multifunctional cytokine that belongs to the Interleukin (IL)-6 subfamily. Since OSM is the most closely related to leukemia inhibitory factor (LIF) structurally, functionally, and genetically among the family members, OSM had

M. Tanaka (✉) · A. Miyajima
Institute of Molecular and Cellular Biosciences,
University of Tokyo,
1-1-1 Yayoi, Bunkyo-ku, 113-0032 Tokyo, Japan
e-mail: tanaka@iam.u-tokyo.ac.jp

M. Tanaka · A. Miyajima
Kanagawa Academy of Science and Technology,
907 Nogawa, Miyamae-ku, 216-0001 Kawasaki, Japan

long been considered as another LIF. In fact, these two cytokines act on a wide variety of cells and elicit diverse overlapping biological responses such as growth regulation, differentiation, gene expression, and cell survival in humans. The functional redundancy can be explained by their shared receptor subunit. However, it has also been recognized that OSM exhibits unique activities that are not shared with LIF and accumulating evidence indicates that OSM is a unique cytokine that plays an important role for various biological systems such as inflammatory response, hematopoiesis, tissue remodeling, and development. This review describes the properties of OSM, including its structure, receptors, signal transduction, biological activities, and gene regulation.

Biochemical and genetic profile of OSM

OSM is a member of the IL-6 subfamily that includes IL-6, IL-11, LIF, ciliary neurotrophic factor (CNTF), cardiotropin-1 (CT-1), and novel neutrophin-1/B cell-stimulating factor-3 (NNT-1/BSF-3) (Kishimoto et al. 1992; Hibi et al. 1996; Senaldi et al. 1999). Human OSM (hOSM) was initially recognized by its activity to inhibit the proliferation of A375 melanoma cells as well as numerous other tumor cells (Zarling et al. 1986). hOSM is a secreted glycoprotein of 28 kDa that was originally isolated from phorbol 12-myristate 13-acetate (PMA)-stimulated human histiocytic lymphoma U937 cells. The hOSM cDNA was isolated from U937 cells that were induced to differentiate into macrophage-like cells by treatment with PMA. The hOSM cDNA encodes a protein of 252 amino acid residues with a signal peptide of 25 amino acid residues (Malik et al. 1989). The C-terminal region of 31 amino acids is removed from the precursor, resulting in the mature form of 196 amino acids (Linsley et al. 1990). The hOSM polypeptide contains five cysteine residues (C6, C49, C80, C127, and C167), which form two intramolecular disulfide bonds, C6-C127 and C49-C167, and forms a secondary structure with four helix bundles, a characteristic of this family of cytokines (Bazan 1991; Hoffman et al. 1996). Genomic DNA analysis revealed that the hOSM gene is located in human chromosome 22q12, and that the coding region is covered by three exons (Malik et al. 1989; Rose et al. 1993). Among the IL-6 cytokine subfamily, OSM and LIF are not only structurally related (Rose et al. 1994), but their genes are also tightly linked on the same chromosomal location, suggesting that the two genes arose by duplication (Rose et al. 1993; Jeffery et al. 1993; Giovannini et al. 1993). Interestingly, mouse OSM (mOSM) shows relatively low identity (48%) with hOSM, while mouse LIF shows high amino acid identity (78%) with human LIF (Yoshimura et al. 1996; Rose et al. 1994). Besides hOSM and mOSM, bovine OSM (bOSM) (Malik et al. 1995) and simian OSM (Rose and Bruce 1991) have been reported to date.

The mOSM cDNA was isolated as an immediate early gene induced by IL-2, IL-3, and erythropoietin (EPO) through the Jak-STAT5 pathway (Yoshimura et al. 1996). The mOSM gene is located in mouse chromosome 11 (Yoshimura et al. 1996). Linkage mapping by interspecific back-cross analysis suggests that OSM and LIF genes were tightly linked within 2.0 cM. The GAS-like sequence, TTCCCAGAA, which is located close to the transcription initiation site, is primarily responsible for induction by IL-2, IL-3, and EPO. mOSM mRNA is abundantly expressed in hematopoietic tissues such as bone marrow, thymus, and spleen. hOSM is secreted from activated T cells, monocytes, and neutrophils (Brown et al. 1987; Malik et al. 1989; Grenier et al. 1999; Hurst et al. 2002), and promoter analysis of hOSM also revealed that GM-CSF-stimulated OSM expression is driven by STAT5 through a *cis*-acting STAT element on the OSM promoter (Ma et al. 1999).

The OSM receptor

It is known that different cytokines often exhibit similar biological activities on the same cell type (functional redundancy). The functional redundancy among the IL-6 family cytokines is now well explained by their receptor structure (Hibi et al. 1996; Heinrich et al. 1998). Functional receptors for this family of cytokines consist of multiple subunits including the common signal transducing subunit, gp130 (Hibi et al. 1990). The receptor complexes for IL-6 and IL-11 consist of a ligand-specific receptor α subunit and gp130. The binding of each cytokine to its specific receptor subunit induces homodimerization of gp130 (Murakami et al. 1993; Yin et al. 1993). The LIF receptor consists of the low-affinity LIF binding protein (LIFRβ) and gp130; LIF binding leads to heterodimerization of LIFRβ and gp130 (Gearing et al. 1991; Gearing et al. 1992). The CNTF receptor is composed of CNTF-specific subunit, LIFRβ and gp130 (Davis et al. 1993). Although OSM is a cytokine that binds to gp130 directly with low affinity, it is not enough to transduce its signals (Gearing et al. 1992; Liu et al. 1992). In human, two types of functional OSM receptor are known: the type I OSM receptor is identical to the high-affinity LIF receptor that consists of gp130 and LIFRβ (Gearing et al. 1992), and the type II OSM receptor consists of gp130 and the OSM-specific receptor subunit (Bruce et al. 1992; Thoma et al. 1994). The cDNA of human OSM-specific receptor β subunit (hOSMRβ) was cloned by polymerase chain reaction (PCR) using human genomic DNA as a template and degenerate oligonucleotides primers designated from a number of homologous regions between gp130, LIFR, and G-CSFR (Mosley et al. 1996). The open reading frame encoded a protein of 979 amino acids, which showed 32.2% and 23.3% identity with hLIFR and gp130, respectively. hOSMRβ was expressed in a wide variety of cell types, including placental cells, smooth muscle cells, skin cells, and many tumor cell lines. While hLIF and hOSM share a number of common biological functions, hOSM displays some specific biological properties that are not shared by hLIF, e.g., growth inhibition of A375 melanoma cells (Bruce et al. 1992), autocrine growth stimulation of AIDS-related Kaposi's sarcoma cells (Miles et al. 1992; Nair et al. 1992; Murakami-Mori et al. 1995), and upregulation of α1-proteinase inhibitor in lung-derived epithelial cells (Cichy et al. 1998). Thus, it seems that many overlapping biological responses between hOSM and hLIF are mediated by the shared type I receptor, i.e., the LIF receptor, while hOSM manifests its specific responses through the type II receptor. Since hOSM and hLIF also display common biological activities on murine cells, e.g., induction of differentiation of M1 mouse myeloid leukemic cells (Bruce et al. 1992) and inhibition of differentiation of mouse embryonic stem (ES) cells (Rose et al. 1994), the effect of hOSM in mouse is believed to mimic mOSM. However, this is not the case. Immediately after the isolation of mOSM cDNA, it was recognized that there are differences in biological activity between human and murine OSM (Ichihara et al. 1997). For example, it was shown that more than 30-fold higher concentration of mOSM was required for the growth inhibition of M1 cells compared with hOSM. Likewise, mOSM was much less potent than hOSM in the inhibition of differentiation of mouse ES cells. In contrast, NIH3T3 mouse embryonic fibroblasts responded to mOSM, but not to mLIF and hOSM. These results indicated that unlike hOSM, mOSM and mLIF did not share the same functional receptor, and mOSM delivered signals only through its

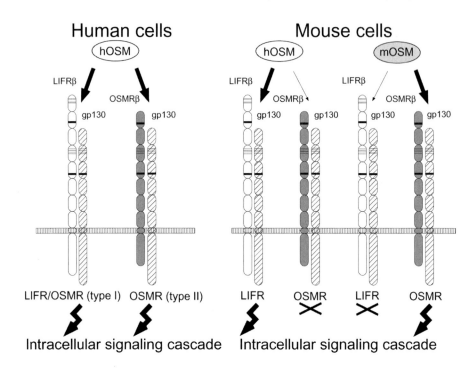

Fig. 1 Formation of the functional receptor complexes for hOSM and mOSM. *Thin horizontal lines* and *broad bars* represent the conserved cysteine residues and WS motifs, respectively. *Thin arrows* represent low-affinity binding of OSM to each receptor component, while *thick arrows* show high-affinity binding. *Zigzag arrows* indicate the relay of intracellular signaling cascades. *X* shows no signaling

specific receptor complex. Molecular cloning of mouse OSMRβ cDNA and reconstitution of the high-affinity functional OSM receptor revealed that mOSM transduces signals through its specific receptor complex composed of gp130 and OSMRβ, but not through the LIF receptor, unless a very high concentration of mOSM is used (Lindberg et al. 1998; Tanaka et al. 1999; Fig. 1). Interestingly, hOSM binds to the mouse LIF receptor and transduces signals, however it fails to transduce signals through the mouse OSM receptor (Richards et al. 1997; Lindberg et al. 1998). Thus, it should be noted that the biological functions of hOSM observed in mouse cells are likely to represent mouse LIF functions. Recently, the expression pattern of OSMRβ during development was reported (Tamura et al. 2002). The expression of mOSMRβ was first detected in aortic endothelial cells of the AGM (aorta-gonad-mesonephros) region at 11.5 days postcoitus (dpc). At 14.5 dpc, mOSMRβ was expressed in the primordia of some organs, including liver, thymus, choroid plexus, and limb. After birth, its gene expression was detectable in other organs, such as lymph node, bone, heart, kidney, small intestine, nasal cavity, and lung.

Signal transduction pathways

OSM activates intracellular signaling cascades through the OSM receptor containing gp130 (Fig. 2). The IL-6 subfamily cytokine receptors do not posses an intrinsic tyrosine

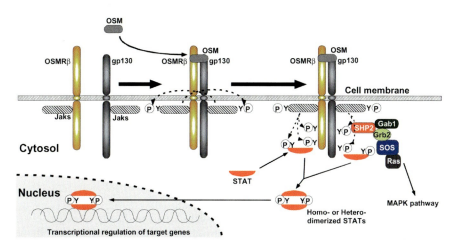

Fig. 2 Signal transduction pathways via the OSM receptor. Binding of OSM to the receptor components induces hetero-dimerization of each subunit, resulting in the reciprocal phosphorylation and activation of Jaks. The activated Jaks phosphorylate tyrosine residues of receptor subunits, providing distinct binding sites for STATs and SHP2. The STATs recruited to the sites are also phosphorylated by Jaks, followed by the homo- or hetero-dimerization of STATs. The dimerized STATs are translocated into the nucleus and regulate transcription of their target genes. The SHP2 recruitment is required for the mitogen-activated protein kinase (*MAPK*) pathway. Y and the *encircled P* represent tyrosine residue and phosphorylation, respectively

kinase but utilize Jaks as the ignition of their signals (Stahl et al. 1995; Gerhartz et al. 1996; Darnell 1997). The first step in the receptor activation is the ligand-induced homo- or hetero-dimerization of signal-transducing receptor subunits. As each signal transducing receptor subunit binds one of the Jaks (Jak1, Jak2, and Tyk2), dimerization of the subunits leads to the reciprocal phosphorylation and activation of Jaks. The activated Jak kinases phosphorylate tyrosine residues in the intracellular domain of the receptor, creating docking sites for STATs as well as various signaling molecules with an SH2 domain. These molecules recruited to the receptors are then activated by Jaks. Phosphorylated STATs, mainly STAT3 and STAT1, then form homo- or hetero-dimers and translocate to the nucleus where they are involved in gene regulation. The IL-6 subfamily cytokines stimulate not only the Jak/STAT signaling pathway but also the Ras/Raf/MAPK signaling pathway (Amaral et al. 1993; Thoma et al. 1994; Boulton et al. 1994). It is known that several adaptor molecules such as SHP2 (Stahl et al. 1995), Grb2 (Neumann et al. 1996), and Gab1 (Takahashi-Tezuka et al. 1998) are involved in this pathway. There are some differences in signal transduction between the type I and type II OSM receptors, e.g., STAT5b is predominantly activated by the OSM-specific type II receptor in the A375 melanoma cell line (Auguste et al. 1997).

Biological activities of OSM

It is known that the IL-6 subfamily cytokines are involved in a variety of biological activities such as inflammation, remodeling of extracellular matrix, hematopoiesis, and modulation of cell growth and differentiation. OSM also exhibits diverse biological activities on a wide variety of cells in vivo and in vitro. As mentioned above, the existence of two functional OSM receptors, type I and type II, provides a molecular basis for the common biological activities between hLIF and hOSM, as well as for hOSM-specific activities. Therefore, it is important to know which receptor is expressed in the biological system of interest. It should be also noted that mOSM uses only the OSM-specific receptor and not the LIF receptor. Since hOSM is able to stimulate murine cells via mLIFR, information about the effect of OSM on distinct species should be interpreted carefully. This chapter mainly describes the biological activities of hOSM on human cells, or mOSM on mouse cells. Moreover, the OSM-specific activities that are not shared by LIF or IL-6 are also discussed.

Growth modulation by OSM

OSM modulates growth of tumor and nontumor cells either positively or negatively depending on the target cells. OSM inhibits the growth of several types of tumor cells such as solid tissue tumor cells, lung cancer cells, melanoma cells, breast cancer cells, and glioma cells (Zarling et al. 1986; Horn et al. 1990; Liu et al. 1997; Halfter et al. 1998). Besides tumor cells, OSM also inhibits proliferation of normal mammary and breast epithelial cells (Liu et al. 1998; Grant et al. 2001). In contrast, OSM stimulates growth of AIDS-related Kaposi's sarcoma cells (Miles et al. 1992; Nair et al. 1992; Murakami-Mori et al. 1995), myeloma cells (Zhang et al. 1994), and plasmacytoma cells (Nishimoto et al. 1994). OSM also stimulates the mitogenesis of normal dermal fibroblasts via mitogen-activated protein kinase (MAPK)-dependent pathway (Ihn and Tamaki 2000). mOSM is known to inhibit growth of a subline of NIH3T3 cells (Hara et al. 1997). mOSM induces differentiation of fetal hepatocytes and downregulates cyclin D expression via STAT3 in an in vitro culture system, while it induces expression of cyclin D in adult hepatocytes (Matsui et al. 2002a). In contrast, mOSM stimulates growth of endothelial-like cells in primary culture of AGM-derived cells and stimulates the development of definitive hematopoiesis in vitro (Mukouyama et al. 1998). mOSM also enhances the proliferation of Sertoli cells derived from neonatal testes (Hara et al. 1998). Although OSM is involved in the growth modulation of various types of cells in vitro, OSMRβ-deficient mice develop normally (M. Tanaka, in preparation). However, it should be noted that no transgenic mouse overexpressing bOSM using the keratin-14 promoter was generated, suggesting that expression of bOSM within developing skin is lethal (Malik et al. 1995). Similarly, the frequency of establishing transgenic mice ubiquitously expressing bOSM using metallothionein promoter was significantly low, suggesting that overexpression of OSM is deleterious during mouse development.

Inflammatory responses by OSM

Production of cytokines and inflammatory proteins and leukocyte adhesion

OSM is secreted from activated T cells and monocytes and plays roles in inflammatory reactions. Acute inflammation is characterized by rapid increase of acute phase proteins (APPs) from the liver. OSM, as well as IL-6, stimulates APPs synthesis in hepatoma and hepatocytes (Richards et al. 1992; Benigni et al. 1996). OSM regulates inflammation not only directly, but also indirectly through the production of other cytokines and their receptors. OSM stimulates the production of IL-6 in cultured endothelial cells (Brown et al.

1991), and the IL-6 receptor in hepatoma HepG2 cells (Cichy et al. 1997). As IL-6 strongly stimulates inflammatory reactions and IL-6-deficient mice exhibit reduced production of APPs and delayed repair of injured liver (Kopf et al. 1994; Fattori et al. 1994; Cressman et al. 1996; Kovalovich et al. 2000), OSM may induce inflammation through the production of IL-6. OSM also modulates other cytokines or chemokines in inflammation, e.g., OSM inhibits the IL-1-induced expression of IL-8 and GM-CSF in synovial and lung fibroblasts (Richards et al. 1996). As neither IL-6 nor LIF displays this activity, the inhibition is likely exerted via the type II OSM receptor. Furthermore, OSM induces mRNA for chemokines, e.g., growth-related oncogene α and β, in human endothelial cells (Modur et al. 1997). OSM induces prolonged expression of P-selectin (Yao et al. 1996) and E-selectin (Modur et al. 1997) in human endothelial cells, which modulate leukocyte adhesion. OSM also induces endothelial cell expression of adhesion molecules such as intercellular adhesion molecule-1 (ICAM-1) and vascular cell adhesion molecule-1 (VCAM-1; Modur et al. 1997). Thus, OSM plays an important role for recruiting leukocytes to inflammatory sites. Similarly, mOSM has been reported to stimulate mouse synovial fibroblasts in vitro and to induce inflammation and destruction in mouse joints in vivo (Langdon et al. 2000).

Remodeling of extracellular matrix

In the inflammatory process, remodeling of the extracellular matrix is important for healing the damaged tissue induced by inflammatory responses. Matrix metalloproteinases (MMPs) are involved in extracellular matrix breakdown, while tissue inhibitors of metalloproteinases (TIMPs) inhibit the action of MMPs (Woessner 1991; Cawston et al. 1994; Cawston et al. 1999). Therefore, the balance between TIMPs and MMPs is important for remodeling of the extracellular matrix. OSM strongly elevates TIMP-1 expression in fibroblast cultures of human lung or synovial origin, whereas IL-6 and LIF marginally increase the expression (Richards et al. 1993). OSM also induces TIMP-1 expression and inhibits IL-1β-induced TIMP-3 in cultured human synovial lining cells (Gatsios et al. 1996). OSM also induces or enhances the expression of MMP-1 and MMP-3 in astrocytes, and that of MMP-1 and MMP-9 in fibroblasts (Korzus et al. 1997). mOSM is also known to stimulate TIMP-1 mRNA in NIH-3T3 mouse embryonic fibroblasts (Richards et al. 1997). Thus, OSM is involved in the wound healing by modulating the balance between TIMPs and MMPs. Moreover, OSM is known to regulate other proteolytic enzymes, such as neutrophil erastase, by expression of proteinase inhibitor. hOSM stimulates α1-antichymotrypsin synthesis in lung-derived epithelial cells (Cichy et al. 1995). Although OSM, LIF, and IL-6 stimulate the production of α1-proteinase inhibitor (α1-PI) in HepG2 cells, only hOSM is able to upregulate levels of α1-PI in lung-derived epithelial cells (Cichy et al. 1998). Interestingly, despite the fact that LIF induces phosphorylation of the LIF receptor, LIF is not able to stimulate α1-PI synthesis in lung epithelial cells. Likewise, IL-6 in combination with the soluble IL-6 receptor induces phosphorylation of gp130 but fails to stimulate expression of α1-PI in lung epithelial cells. Thus, OSM-specific signaling plays an important role in lung inflammation.

Roles of OSM in liver development and regeneration

As previously described, OSM stimulates hepatocytes and induces APPs. In addition to APPs, OSM modulates the expression of other molecules in liver cells, e.g., OSM upregu-

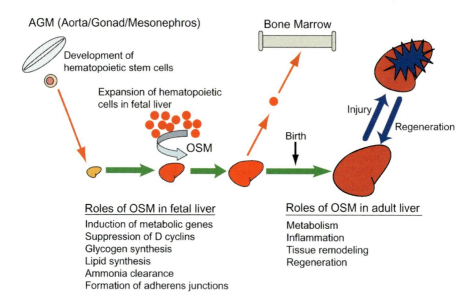

Fig. 3 The actions of OSM on liver. Hematopoietic stem cells (*HSCs*) developed in the AGM region migrate to the fetal liver, and expand in hepatic microenvironment. OSM produced by hematopoietic cells promote differentiation of hepatic cells, which is accompanied by functional and morphological maturation. As the fetal liver matures, it gradually loses hematopoietic potential, and HSCs relocate to the bone marrow

lates expression of low density lipoprotein receptors (Grove et al. 1991) and protein S (Hooper et al. 1995) in HepG2 cells, whereas it downregulates the expression of some cytochrome P450 (CYP) isozymes such as CYP1A2 and CYP3A4 in human hepatocytes at the transcriptional level (Guillen et al. 1998). OSM is also known to be upregulated in cirrhotic human liver (Levy et al. 2000). Consistently, OSM increases collagen production by hepatic stellate cells without induction of collagen mRNA, possibly by posttranscriptional mechanism. OSM also affects fetal liver development (reviewed by Miyajima et al. 2000; Kinoshita and Miyajima 2002). In embryonic days (E) 14.5 fetal liver culture, mOSM stimulates the functional maturation of fetal hepatocytes such as expression of hepatic differentiation markers, intracellular accumulation of glycogen, lipid synthesis, and clearance of ammonia (Kamiya et al. 1999; Kamiya et al. 2002; Kojima et al. 2000; Sakai et al. 2002). STAT3 is an essential signaling component for OSM-induced hepatic development, while activation of Ras appears to negatively regulate the process (Ito et al. 2000). Furthermore, mOSM enhances E-cadherin-based adherens junction formation of hepatocytes via K-Ras and induces morphological maturation (Matsui et al. 2002b). OSM downregulates expression of D1 and D2 cyclins in fetal hepatocytes through STAT3 (Matsui et al. 2002a), while OSM as well as IL-6 upregulates expression of D cyclins in adult hepatocytes in regenerating liver (K. Nakamura, in preparation). mOSM is expressed in CD45-positive hematopoietic cells in fetal liver, whereas OSMRβ is predominantly detected in hepatocytes, suggesting that OSM induces development of hepatocytes in a paracrine manner (Kamiya et al. 1999; Kinoshita et al. 1999). The overview of OSM actions on liver is shown in Fig. 3.

Hematopoiesis

It is known that OSM mRNA is abundant in hematopoietic tissues such as bone marrow, thymus and spleen (Yoshimura et al. 1996). In addition, OSM is expressed in the mouse AGM region at 11.5 dpc, where long-term repopulating hematopoietic stem cells (LTR-HSCs) first arise (Mukouyama et al. 1998). Similarly, mOSMRβ expression is detected in aortic endothelial cells of the AGM region at 11.5 dpc (Tamura et al. 2002) and adult thymus and spleen (Tanaka et al. 1999), implying that OSM plays a role in hematopoiesis. In fact, it has been reported that OSM stimulates the formation of endothelial cell clusters in the primary cultures of mouse AGM and induces the expansion of multipotential hematopoietic progenitors (Mukouyama et al. 1998). Furthermore, coculture of E11.5 AGM cells and E14.5 fetal liver cells in the presence of mOSM resulted in the expansion of LTR-HSCs (Takeuchi et al. 2002). Not only fetal hematopoiesis, but also adult hematopoiesis is modulated by OSM. Transgenic overexpression of bOSM in mice caused hematopoietic abnormalities such as splenomegaly and expansion of megakaryocytes in bone marrow (Malik et al. 1995). It is also known that hOSM, as well as the other members of the IL-6 family, IL-6 or LIF, can markedly enhance megakaryocytic colony formation from murine marrow cells in combination with murine IL-3 in vitro (Ishibashi et al. 1989; Wallace et al. 1995). Similarly, the administration of hOSM in normal mice augments platelet counts. Transplantation of mouse bone marrow cells constitutively expressing OSM induces a myeloproliferative phenotype that partially recapitulates the TEL/Jak2 disease (Schwaller et al. 2000). This result is consistent with the observation that OSM expression is induced by cytokines that activate Jak2 and STAT5. Recently, it was reported that the number of hematopoietic progenitor cells in bone marrow of STAT4-deficient mice was significantly reduced, but recovered with T-cell-specific transgenic expression of STAT4, suggesting that T cells, specifically Th1 cells, play an important role in hematopoiesis (Broxmeyer et al. 2002). Moreover, injection of the Th1 cytokine, OSM, but not other cytokines, into STAT4-deficient mice recovered progenitors to wild-type levels, suggesting that OSM is a potential mediator between T cells and hematopoietic progenitors. Furthermore, OSM is also involved in lymphopoiesis. Transgenic mouse expressing OSM (either human, bovine, or mouse) by the lck promoter exhibited dramatic accumulation of immature and mature T cells in lymph nodes, suggesting that overexpression of OSM in early T cells results in an extrathymic T-cell development in lymph nodes (Malik et al. 1995; Clegg et al. 1996; Clegg et al. 1999). It is also reported that the administration of OSM, IL-6, and LIF into mice induces acute thymic atrophy with a reduced number of thymocytes (Sempowski et al. 2000). Thus, OSM regulates hematopoiesis at various hematopoietic organs and stages. In fact, targeted disruption of OSMRβ resulted in altered hematopoiesis (M. Tanaka, in preparation). The numbers of peripheral erythrocytes and platelets in OSMRβ-deficient mice were significantly reduced compared with wild-type mice. Consistent with this, progenitors of erythroid and megakaryocyte lineages were reduced in mutant bone marrow. Our results suggest that OSM affects hematopoietic microenvironments.

Future directions

Because OSM and LIF exhibit a number of common biological, biochemical and genetical characteristics, OSM was considered to be another LIF. However, it has become clear that OSM is a unique cytokine with multiple functions as shown by its unique biological activities in liver, lung, testes, and hematopoietic organs. Like many IL-6 family members, OSM induces a variety of biological functions via gp130 through STAT3 and Ras/MAPK pathways. However, activation of gp130 by IL-6 in combination with the soluble IL-6 receptor does not always mimic OSM activity, suggesting a unique role for OSMRβ in signaling. While the OSMRβ-deficient mice we have generated develop normally, alterations in hematopoiesis as well as liver regeneration have been noticed. Therefore, detailed analyses of this mutant mouse will uncover its unique functions in various aspects of life and also provide useful information as to the development of drugs for diseases such as inflammation.

References

Amaral MC, Miles S, Kumar G, Nel AE (1993) Oncostatin-M stimulates tyrosine protein phosphorylation in parallel with the activation of p42MAPK/ERK-2 in Kaposi's cells. Evidence that this pathway is important in Kaposi cell growth. J Clin Invest 92:848–857

Auguste P, Guillet C, Fourcin M, Olivier C, Veziers J, Pouplard-Barthelaix A, Gascan H (1997) Signaling of type II oncostatin M receptor. J Biol Chem 272:15760–15764

Bazan JF (1991) Neuropoietic cytokines in the hematopoietic fold. Neuron 7:197–208

Benigni F, Fantuzzi G, Sacco S, Sironi M, Pozzi P, Dinarello CA, Sipe JD, Poli V, Cappelletti M, Paonessa G, Pennica D, Panayotatos N, Ghezzi P (1996) Six different cytokines that share GP130 as a receptor subunit, induce serum amyloid A and potentiate the induction of interleukin-6 and the activation of the hypothalamus-pituitary-adrenal axis by interleukin-1. Blood 87:1851–1854

Boulton TG, Stahl N, Yancopoulos GD (1994) Ciliary neurotrophic factor/leukemia inhibitory factor/interleukin 6/oncostatin M family of cytokines induces tyrosine phosphorylation of a common set of proteins overlapping those induced by other cytokines and growth factors. J Biol Chem 269:11648–11655

Brown TJ, Lioubin MN, Marquardt H (1987) Purification and characterization of cytostatic lymphokines produced by activated human T lymphocytes. Synergistic antiproliferative activity of transforming growth factor β 1, interferon-γ, and oncostatin M for human melanoma cells. J Immunol 139:2977–2983

Brown TJ, Rowe JM, Liu JW, Shoyab M (1991) Regulation of IL-6 expression by oncostatin M. J Immunol 147:2175–2180

Broxmeyer HE, Bruns HA, Zhang S, Cooper S, Hangoc G, McKenzie AN, Dent AL, Schindler U, Naeger LK, Hoey T, Kaplan MH (2002) Th1 cells regulate hematopoietic progenitor cell homeostasis by production of oncostatin M. Immunity 16:815–825

Bruce AG, Hoggatt IH, Rose TM (1992) Oncostatin M is a differentiation factor for myeloid leukemia cells. J Immunol 149:1271–1275

Cawston T, Plumpton T, Curry V, Ellis A, Powell L (1994) Role of TIMP and MMP inhibition in preventing connective tissue breakdown. Ann N Y Acad Sci 732:75–83

Cawston T, Billington C, Cleaver C, Elliott S, Hui W, Koshy P, Shingleton B, Rowan A (1999) The regulation of MMPs and TIMPs in cartilage turnover. Ann N Y Acad Sci 878:120–129

Cichy J, Potempa J, Chawla RK, Travis J (1995) Stimulatory effect of inflammatory cytokines on α1-antichymotrypsin expression in human lung-derived epithelial cells. J Clin Invest 95:2729–2733

Cichy J, Rose-John S, Potempa J, Pryjma J, Travis J (1997) Oncostatin M stimulates the expression and release of the IL-6 receptor in human hepatoma HepG2 cells. J Immunol 159:5648–5653

Cichy J, Rose-John S, Travis J (1998) Oncostatin M, leukaemia-inhibitory factor and interleukin 6 trigger different effects on α1-proteinase inhibitor synthesis in human lung-derived epithelial cells. Biochem J 329:335–339

Clegg CH, Rulffes JT, Wallace PM, Haugen HS (1996) Regulation of an extrathymic T-cell development pathway by oncostatin M. Nature 384:261–263

Clegg CH, Haugen HS, Rulffes JT, Friend SL, Farr AG (1999) Oncostatin M transforms lymphoid tissue function in transgenic mice by stimulating lymph node T-cell development and thymus autoantibody production. Exp Hematol 27:712–725

Cressman DE, Greenbaum LE, DeAngelis RA, Ciliberto G, Furth EE, Poli V, Taub R (1996) Liver failure and defective hepatocyte regeneration in interleukin-6-deficient mice. Science 274:1379–1383

Darnell JE Jr. (1997) STATs and gene regulation. Science 277:1630–1635

Davis S, Aldrich TH, Stahl N, Pan L, Taga T, Kishimoto T, Ip NY, Yancopoulos GD (1993) LIFR β and gp130 as heterodimerizing signal transducers of the tripartite CNTF receptor. Science 260:1805–1808

Fattori E, Cappelletti M, Costa P, Sellitto C, Cantoni L, Carelli M, Faggioni R, Fantuzzi G, Ghezzi P, Poli V (1994) Defective inflammatory response in interleukin 6-deficient mice. J Exp Med 180:1243–1250

Gatsios P, Haubeck HD, Van de Leur E, Frisch W, Apte SS, Greiling H, Heinrich PC, Graeve L (1996) Oncostatin M differentially regulates tissue inhibitors of metalloproteinases TIMP-1 and TIMP-3 gene expression in human synovial lining cells. Eur J Biochem 241:56–63

Gearing DP, Thut CJ, VandeBos T, Gimpel SD, Delaney PB, King J, Price V, Cosman D, Beckmann MP (1991) Leukemia inhibitory factor receptor is structurally related to the IL-6 signal transducer, gp130. EMBO J 10:2839–2848

Gearing DP, Comeau MR, Friend DJ, Gimpel SD, Thut CJ, McGourty J, Brasher KK, King JA, Gillis S, Mosley B (1992) The IL-6 signal transducer, gp130: an oncostatin M receptor and affinity converter for the LIF receptor. Science 255:1434–1437

Gerhartz C, Heesel B, Sasse J, Hemmann U, Landgraf C, Schneider-Mergener J, Horn F, Heinrich PC, Graeve L (1996) Differential activation of acute phase response factor/STAT3 and STAT1 via the cytoplasmic domain of the interleukin 6 signal transducer gp130. I. Definition of a novel phosphotyrosine motif mediating STAT1 activation. J Biol Chem 271:12991–12998

Giovannini M, Djabali M, McElligott D, Selleri L, Evans GA (1993) Tandem linkage of genes coding for leukemia inhibitory factor (LIF) and oncostatin M (OSM) on human chromosome 22. Cytogenet Cell Genet 64:240–244

Grant SL, Douglas AM, Goss GA, Begley CG (2001) Oncostatin M and leukemia inhibitory factor regulate the growth of normal human breast epithelial cells. Growth Factors 19:153–162

Grenier A, Dehoux M, Boutten A, Arce-Vicioso M, Durand G, Gougerot-Pocidalo MA, Chollet-Martin S (1999) Oncostatin M production and regulation by human polymorphonuclear neutrophils. Blood 93:1413–1421

Grove RI, Mazzucco CE, Radka SF, Shoyab M, Kiener PA (1991) Oncostatin M up-regulates low density lipoprotein receptors in HepG2 cells by a novel mechanism. J Biol Chem 266:18194–18199

Guillen MI, Donato MT, Jover R, Castell JV, Fabra R, Trullenque R, Gomez-Lechon MJ (1998) Oncostatin M down-regulates basal and induced cytochromes P450 in human hepatocytes. J Pharmacol Exp Ther 285:127–134

Halfter H, Lotfi R, Westermann R, Young P, Ringelstein EB, Stogbauer FT (1998) Inhibition of growth and induction of differentiation of glioma cell lines by oncostatin M (OSM). Growth Factors 15:135–147

Hara T, Ichihara M, Yoshimura A, Miyajima A (1997) Cloning and biological activity of murine oncostatin M. Leukemia 11 3:449–450

Hara T, Tamura K, de Miguel MP, Mukouyama Y, Kim H, Kogo H, Donovan PJ, Miyajima A (1998) Distinct roles of oncostatin M and leukemia inhibitory factor in the development of primordial germ cells and sertoli cells in mice. Dev Biol 201:144–153

Heinrich PC, Behrmann I, Muller-Newen G, Schaper F, Graeve L (1998) Interleukin-6-type cytokine signalling through the gp130/Jak/STAT pathway. Biochem J 334:297–314

Hibi M, Murakami M, Saito M, Hirano T, Taga T, Kishimoto T (1990) Molecular cloning and expression of an IL-6 signal transducer, gp130. Cell 63:1149–1157

Hibi M, Nakajima K, Hirano T (1996) IL-6 cytokine family and signal transduction: a model of the cytokine system. J Mol Med 74:1–12

Hoffman RC, Moy FJ, Price V, Richardson J, Kaubisch D, Frieden EA, Krakover JD, Castner BJ, King J, March CJ, Powers R (1996) Resonance assignments for oncostatin M, a 24-kDa α-helical protein. J Biomol NMR 7:273–282

Hooper WC, Phillips DJ, Ribeiro M, Benson J, Evatt BL (1995) IL-6 upregulates protein S expression in the HepG-2 hepatoma cells. Thromb Haemost 73:819–824

Horn D, Fitzpatrick WC, Gompper PT, Ochs V, Bolton-Hansen M, Zarling J, Malik N, Todaro GJ, Linsley PS (1990) Regulation of cell growth by recombinant oncostatin M. Growth Factors 2:157–165

Hurst SM, McLoughlin RM, Monslow J, Owens S, Morgan L, Fuller GM, Topley N, Jones SA (2002) Secretion of oncostatin m by infiltrating neutrophils: regulation of IL-6 and chemokine expression in human mesothelial cells. J Immunol 169:5244–5251

Ichihara M, Hara T, Kim H, Murate T, Miyajima A (1997) Oncostatin M and leukemia inhibitory factor do not use the same functional receptor in mice. Blood 90:165–173

Ihn H, Tamaki K (2000) Oncostatin M stimulates the growth of dermal fibroblasts via a mitogen-activated protein kinase-dependent pathway. J Immunol 165:2149–2155

Ishibashi T, Kimura H, Uchida T, Kariyone S, Friese P, Burstein SA (1989) Human interleukin 6 is a direct promoter of maturation of megakaryocytes in vitro. Proc Natl Acad Sci USA 86:5953–5957

Ito Y, Matsui T, Kamiya A, Kinoshita T, Miyajima A (2000) Retroviral gene transfer of signaling molecules into murine fetal hepatocytes defines distinct roles for the STAT3 and ras pathways during hepatic development. Hepatology 32:1370–1376

Jeffery E, Price V, Gearing DP (1993) Close proximity of the genes for leukemia inhibitory factor and oncostatin M. Cytokine 5:107–111

Kamiya A, Kinoshita T, Ito Y, Matsui T, Morikawa Y, Senba E, Nakashima K, Taga T, Yoshida K, Kishimoto T, Miyajima A (1999) Fetal liver development requires a paracrine action of oncostatin M through the gp130 signal transducer. EMBO J 18:2127–2136

Kamiya A, Kojima N, Kinoshita T, Sakai Y, Miyajima A (2002) Maturation of fetal hepatocytes in vitro by extracellular matrices and oncostatin M: induction of tryptophan oxygenase. Hepatology 35:1351–1359

Kinoshita T, Sekiguchi T, Xu MJ, Ito Y, Kamiya A, Tsuji K, Nakahata T, Miyajima A (1999) Hepatic differentiation induced by oncostatin M attenuates fetal liver hematopoiesis. Proc Natl Acad Sci USA 96:7265–7270

Kinoshita T, Miyajima A (2002) Cytokine regulation of liver development. Biochimica Biophysica Acta 1592:303–312

Kishimoto T, Akira S, Taga T (1992) Interleukin-6 and its receptor: a paradigm for cytokines. Science 258:593–597

Kojima N, Kinoshita T, Kamiya A, Nakamura K, Nakashima K, Taga T, Miyajima A (2000) Cell density-dependent regulation of hepatic development by a gp130-independent pathway. Biochem Biophys Res Commun 277:152–158

Kopf M, Baumann H, Freer G, Freudenberg M, Lamers M, Kishimoto T, Zinkernagel R, Bluethmann H, Kohler G (1994) Impaired immune and acute-phase responses in interleukin-6-deficient mice. Nature 368:339–342

Korzus E, Nagase H, Rydell R, Travis J (1997) The mitogen-activated protein kinase and JAK-STAT signaling pathways are required for an oncostatin M-responsive element-mediated activation of matrix metalloproteinase 1 gene expression. J Biol Chem 272:1188–1196

Kovalovich K, DeAngelis RA, Li W, Furth EE, Ciliberto G, Taub R (2000) Increased toxin-induced liver injury and fibrosis in interleukin-6-deficient mice. Hepatology 31:149–159

Langdon C, Kerr C, Hassen M, Hara T, Arsenault AL, Richards CD (2000) Murine oncostatin M stimulates mouse synovial fibroblasts in vitro and induces inflammation and destruction in mouse joints in vivo. Am J Pathol 157:1187–1196

Levy MT, Trojanowska M, Reuben A (2000) Oncostatin M: a cytokine upregulated in human cirrhosis, increases collagen production by human hepatic stellate cells. J Hepatol 32:218–226

Lindberg RA, Juan TS, Welcher AA, Sun Y, Cupples R, Guthrie B, Fletcher FA (1998) Cloning and characterization of a specific receptor for mouse oncostatin M. Mol Cell Biol 18:3357–3367

Linsley PS, Kallestad J, Ochs V, Neubauer M (1990) Cleavage of a hydrophilic C-terminal domain increases growth-inhibitory activity of oncostatin M. Mol Cell Biol 10:1882–1890

Liu J, Modrell B, Aruffo A, Marken JS, Taga T, Yasukawa K, Murakami M, Kishimoto T, Shoyab M (1992) Interleukin-6 signal transducer gp130 mediates oncostatin M signaling. J Biol Chem 267:16763–16766

Liu J, Spence MJ, Wallace PM, Forcier K, Hellstrom I, Vestal RE (1997) Oncostatin M-specific receptor mediates inhibition of breast cancer cell growth and downregulation of the c-myc proto-oncogene. Cell Growth Differ 8:667–676

Liu J, Hadjokas N, Mosley B, Estrov Z, Spence MJ, Vestal RE (1998) Oncostatin M-specific receptor expression and function in regulating cell proliferation of normal and malignant mammary epithelial cells. Cytokine 10:295–302

Ma Y, Streiff RJ, Liu J, Spence MJ, Vestal RE (1999) Cloning and characterization of human oncostatin M promoter. Nucleic Acids Res 27:4649–4657

Malik N, Kallestad JC, Gunderson NL, Austin SD, Neubauer MG, Ochs V, Marquardt H, Zarling JM, Shoyab M, Wei CM (1989) Molecular cloning, sequence analysis, and functional expression of a novel growth regulator, oncostatin M. Mol Cell Biol 9:2847–2853

Malik N, Haugen HS, Modrell B, Shoyab M, Clegg CH (1995) Developmental abnormalities in mice transgenic for bovine oncostatin M. Mol Cell Biol 15:2349–2358

Matsui T, Kinoshita T, Hirano T, Yokota T, Miyajima A (2002a) STAT3 down-regulates the expression of cyclin D during liver development. J Biol Chem 277:36167–36173

Matsui T, Kinoshita T, Morikawa Y, Tohya K, Katsuki M, Ito Y, Kamiya A, Miyajima A (2002b) K-Ras mediates cytokine-induced formation of E-cadherin-based adherens junctions during liver development. EMBO J 21:1021–1030

Miles SA, Martinez-Maza O, Rezai A, Magpantay L, Kishimoto T, Nakamura S, Radka SF, Linsley PS (1992) Oncostatin M as a potent mitogen for AIDS-Kaposi's sarcoma-derived cells. Science 255:1432–1434

Miyajima A, Kinoshita T, Tanaka M, Kamiya A, Mukouyama Y, Hara T (2000) Role of Oncostatin M in hematopoiesis and liver development. Cytokine Growth Factor Rev 11:177–183

Modur V, Feldhaus MJ, Weyrich AS, Jicha DL, Prescott SM, Zimmerman GA, McIntyre TM (1997) Oncostatin M is a proinflammatory mediator. In vivo effects correlate with endothelial cell expression of inflammatory cytokines and adhesion molecules. J Clin Invest 100:158–168

Mosley B, De Imus C, Friend D, Boiani N, Thoma B, Park LS, Cosman D (1996) Dual oncostatin M (OSM) receptors. Cloning and characterization of an alternative signaling subunit conferring OSM-specific receptor activation. J Biol Chem 271:32635–32643

Mukouyama Y, Hara T, Xu M, Tamura K, Donovan PJ, Kim H, Kogo H, Tsuji K, Nakahata T, Miyajima A (1998) In vitro expansion of murine multipotential hematopoietic progenitors from the embryonic aorta-gonad-mesonephros region. Immunity 8:105–114

Murakami M, Hibi M, Nakagawa N, Nakagawa T, Yasukawa K, Yamanishi K, Taga T, Kishimoto T (1993) IL-6-induced homodimerization of gp130 and associated activation of a tyrosine kinase. Science 260:1808–1810

Murakami-Mori K, Taga T, Kishimoto T, Nakamura S (1995) AIDS-associated Kaposi's sarcoma (KS) cells express oncostatin M (OM)-specific receptor but not leukemia inhibitory factor/OM receptor or interleukin-6 receptor. Complete block of OM-induced KS cell growth and OM binding by anti-gp130 antibodies. J Clin Invest 96:1319–1327

Nair BC, DeVico AL, Nakamura S, Copeland TD, Chen Y, Patel A, O'Neil T, Oroszlan S, Gallo RC, Sarngadharan MG (1992) Identification of a major growth factor for AIDS-Kaposi's sarcoma cells as oncostatin M. Science 255:1430–1432

Neumann C, Zehentmaier G, Danhauser-Riedl S, Emmerich B, Hallek M (1996) Interleukin-6 induces tyrosine phosphorylation of the Ras activating protein Shc, and its complex formation with Grb2 in the human multiple myeloma cell line LP-1. Eur J Immunol 26:379–384

Nishimoto N, Ogata A, Shima Y, Tani Y, Ogawa H, Nakagawa M, Sugiyama H, Yoshizaki K, Kishimoto T (1994) Oncostatin M, leukemia inhibitory factor, and interleukin 6 induce the proliferation of human plasmacytoma cells via the common signal transducer, gp130. J Exp Med 179:1343–1347

Richards CD, Brown TJ, Shoyab M, Baumann H, Gauldie J (1992) Recombinant oncostatin M stimulates the production of acute phase proteins in HepG2 cells and rat primary hepatocytes in vitro. J Immunol 148:1731–1736

Richards CD, Shoyab M, Brown TJ, Gauldie J (1993) Selective regulation of metalloproteinase inhibitor (TIMP-1) by oncostatin M in fibroblasts in culture. J Immunol 150:5596–5603

Richards CD, Langdon C, Botelho F, Brown TJ, Agro A (1996) Oncostatin M inhibits IL-1-induced expression of IL-8 and granulocyte-macrophage colony-stimulating factor by synovial and lung fibroblasts. J Immunol 156:343–349

Richards CD, Kerr C, Tanaka M, Hara T, Miyajima A, Pennica D, Botelho F, Langdon CM (1997) Regulation of tissue inhibitor of metalloproteinase-1 in fibroblasts and acute phase proteins in hepatocytes in vitro by mouse oncostatin M, cardiotrophin-1, and IL-6. J Immunol 159:2431–2437

Rose TM, Bruce AG (1991) Oncostatin M is a member of a cytokine family that includes leukemia-inhibitory factor, granulocyte colony-stimulating factor, and interleukin 6. Proc Natl Acad Sci USA 88:8641–8645

Rose TM, Lagrou MJ, Fransson I, Werelius B, Delattre O, Thomas G, de Jong PJ, Todaro GJ, Dumanski JP (1993) The genes for oncostatin M (OSM) and leukemia inhibitory factor (LIF) are tightly linked on human chromosome 22. Genomics 17:136–140

Rose TM, Weiford DM, Gunderson NL, Bruce AG (1994) Oncostatin M (OSM) inhibits the differentiation of pluripotent embryonic stem cells in vitro. Cytokine 6:48–54

Sakai Y, Jiang J, Kojima N, Kinoshita T, Miyajima A (2002) Enhanced in vitro maturation of fetal mouse liver cells with oncostatin M, nicotinamide, and dimethyl sulfoxide. Cell Transplant 11:435–441

Schwaller J, Parganas E, Wang D, Cain D, Aster JC, Williams IR, Lee CK, Gerthner R, Kitamura T, Frantsve J, Anastasiadou E, Loh ML, Levy DE, Ihle JN, Gilliland DG (2000) Stat5 is essential for the myelo- and lymphoproliferative disease induced by TEL/JAK2. Mol Cell 6:693–704

Sempowski GD, Hale LP, Sundy JS, Massey JM, Koup RA, Douek DC, Patel DD, Haynes BF (2000) Leukemia inhibitory factor, oncostatin M, IL-6, and stem cell factor mRNA expression in human thymus increases with age and is associated with thymic atrophy. J Immunol 164:2180–2187

Senaldi G, Varnum BC, Sarmiento U, Starnes C, Lile J, Scully S, Guo J, Elliott G, McNinch J, Shaklee CL, Freeman D, Manu F, Simonet WS, Boone T, Chang MS (1999) Novel neurotrophin-1/B cell-stimulating factor-3: a cytokine of the IL-6 family. Proc Natl Acad Sci USA 96:11458–11463

Stahl N, Farruggella TJ, Boulton TG, Zhong Z, Darnell JE, Jr. and Yancopoulos GD (1995) Choice of STATs and other substrates specified by modular tyrosine-based motifs in cytokine receptors. Science 267:1349–1353

Takahashi-Tezuka M, Yoshida Y, Fukada T, Ohtani T, Yamanaka Y, Nishida K, Nakajima K, Hibi M, Hirano T (1998) Gab1 acts as an adapter molecule linking the cytokine receptor gp130 to ERK mitogen-activated protein kinase. Mol Cell Biol 18:4109–4117

Takeuchi M, Sekiguchi T, Hara T, Kinoshita T, Miyajima A (2002) Cultivation of aorta-gonad-mesonephros-derived hematopoietic stem cells in the fetal liver microenvironment amplifies long-term repopulating activity and enhances engraftment to the bone marrow. Blood 99:1190–1196

Tamura S, Morikawa Y, Tanaka M, Miyajima A, Senba E (2002) Developmental expression pattern of oncostatin M receptor β in mice. Mech Dev 115:127–131

Tanaka M, Hara T, Copeland NG, Gilbert DJ, Jenkins NA, Miyajima A (1999) Reconstitution of the functional mouse oncostatin M (OSM) receptor: molecular cloning of the mouse OSM receptor β subunit. Blood 93:804–815

Thoma B, Bird TA, Friend DJ, Gearing DP, Dower SK (1994) Oncostatin M and leukemia inhibitory factor trigger overlapping and different signals through partially shared receptor complexes. J Biol Chem 269:6215–6222

Wallace PM, MacMaster JF, Rillema JR, Peng J, Burstein SA, Shoyab M (1995) Thrombocytopoietic properties of oncostatin M. Blood 86:1310–1315

Woessner JF, Jr. (1991) Matrix metalloproteinases and their inhibitors in connective tissue remodeling. FASEB J 5:2145–2154

Yao L, Pan J, Setiadi H, Patel KD, McEver RP (1996) Interleukin 4 or oncostatin M induces a prolonged increase in P-selectin mRNA and protein in human endothelial cells. J Exp Med 184:81–92

Yin T, Taga T, Tsang ML, Yasukawa K, Kishimoto T, Yang YC (1993) Involvement of IL-6 signal transducer gp130 in IL-11-mediated signal transduction. J Immunol 151:2555–2561

Yoshimura A, Ichihara M, Kinjyo I, Moriyama M, Copeland NG, Gilbert DJ, Jenkins NA, Hara T, Miyajima A (1996) Mouse oncostatin M: an immediate early gene induced by multiple cytokines through the JAK-STAT5 pathway. EMBO J 15:1055–1063

Zarling JM, Shoyab M, Marquardt H, Hanson MB, Lioubin MN, Todaro GJ (1986) Oncostatin M: a growth regulator produced by differentiated histiocytic lymphoma cells. Proc Natl Acad Sci USA 83:9739–9743

Zhang XG, Gu JJ, Lu ZY, Yasukawa K, Yancopoulos GD, Turner K, Shoyab M, Taga T, Kishimoto T, Bataille R (1994) Ciliary neurotropic factor, interleukin 11, leukemia inhibitory factor, and oncostatin M are growth factors for human myeloma cell lines using the interleukin 6 signal transducer gp130. J Exp Med 179:1337–1342

G. J. M. van de Geijn · L. H. J. Aarts · S. J. Erkeland · J. M. Prasher · I. P. Touw

Granulocyte colony-stimulating factor and its receptor in normal hematopoietic cell development and myeloid disease

Published online: 29 March 2003
© Springer-Verlag 2003

Abstract Hematopoiesis, the process of blood cell formation, is orchestrated by cytokines and growth factors that stimulate the expansion of different progenitor cell subsets and regulate their survival and differentiation into mature blood cells. Granulocyte colony-stimulating factor (G-CSF) is the major hematopoietic growth factor involved in the control of neutrophil development. G-CSF is now applied on a routine basis in the clinic for treatment of congenital and acquired neutropenias. G-CSF activates a receptor of the hematopoietin receptor superfamily, the G-CSF receptor (G-CSF-R), which subsequently triggers multiple signaling mechanisms. Here we review how these mechanisms contribute to the specific responses of hematopoietic cells to G-CSF and how perturbations in the function of the G-CSF-R are implicated in various types of myeloid disease.

G-CSF

The cloning and functional characterization of hematopoietic growth factors and their cell surface receptors represent milestones in understanding the molecular control of blood cell development (Nagata et al. 1986; Souza et al. 1986; D'Andrea et al. 1989; Metcalf 1989; Larsen et al. 1990; Metcalf 1991; Lyman 1995). In addition, these developments have had a profound impact on clinical hematology, most notably through the introduction of hematopoietic growth factor-based therapies. Granulocyte-colony stimulating factor (G-CSF) is a member of the cytokine class I superfamily, structurally characterized by four antiparallel α-helices (Wells and de Vos 1996). G-CSF supports the proliferation, survival, and differentiation of neutrophilic progenitor cells in vitro and provides nonredundant signals for maintenance of steady-state neutrophil levels in vivo (Demetri and Griffin 1991; Lieschke et al. 1994; Berliner et al. 1995; Liu et al. 1996; Lieschke 1997). Typically, G-CSF-deficient (*gcsf-/-*) or G-CSF-receptor-deficient (*gcsfr-/-*) mice manifest a selective neutrope-

G. J. M. van de Geijn · L. H. J. Aarts · S. J. Erkeland · J. M. Prasher · I. P. Touw (✉)
Department of Hematology, Erasmus University Medical Center,
P.O. Box 1738, 3000 DR Rotterdam, The Netherlands
e-mail: i.touw@erasmusmc.nl · Tel.: +31-10-4087837 · Fax: +31-10-4089470

nia, with blood neutrophil levels at 15%–30% of those in wild-type (wt) littermates. The number of myeloid progenitor cells in the bone marrow of these mice is also significantly decreased (Lieschke et al. 1994; Liu et al. 1996; Hermans et al. 2002). Experiments with G-CSF- or G-CSF-R-deficient mice infected with *Listeria monocytogenes* have established that G-CSF signaling is also required for "reactive" or "emergency" granulopoiesis in response to bacterial infections (Lieschke et al. 1994; Zhan et al. 1998). In addition, G-CSF enhances neutrophil effector functions, such as superoxide anion generation, release of arachidonic acid and production of leukocyte alkaline phosphatase and myeloperoxidase by mature neutrophils (Morishita et al. 1987; Sato et al. 1988; Avalos et al. 1990).

The clinical application of G-CSF has been particularly beneficial in the treatment of various forms of neutropenia. For example, this is the case for severe congenital neutropenia (SCN), a disease characterized by a myeloid maturation arrest in the bone marrow leading to a drastic reduction in peripheral neutrophil levels and susceptibility to opportunistic bacterial infections that can be fatal. G-CSF treatment ameliorates the neutropenia and associated infections in a large majority of cases (Welte et al. 1990; Dale et al. 1993; Welte and Boxer 1997). Another major and initially unexpected benefit of G-CSF is its ability to induce the egress of hematopoietic stem and progenitor cells from the bone marrow into the peripheral blood. This has resulted in utilization of G-CSF in the mobilization and isolation of peripheral hematopoietic stem cells for transplantation purposes (Molineux et al. 1990). The mechanism by which G-CSF mobilizes these cells into the periphery is not fully understood but is thought to involve multiple effector pathways, including proteolytic enzyme release, activation of chemokine receptors, and modulation of adhesion molecules (Lapidot and Petit 2002; Thomas et al. 2002). G-CSF also induces the mobilization of neutrophils from the bone marrow, probably via similar mechanisms (Semerad et al. 2002).

G-CSF receptor

The G-CSF-R is a member of the now well-characterized hematopoietin receptor superfamily (Bazan 1990; Cosman 1993). This family is structurally characterized by four highly conserved cysteine residues and a tryptophan-serine repeat (WSXWS) in the extracellular domain. Both motifs are located within the so-called cytokine receptor homology (CRH) region. Murine and human G-CSF receptors are single transmembrane proteins of 812 and 813 amino acid residues, respectively, with 62.5% homology at the amino acid level (Fukunaga et al. 1990). The extracellular domain of the G-CSF-R contains 603 amino acids and includes an immunoglobulin-like module, the CRH domain, and three fibronectin (FN) type III modules.

The CRH domain is composed of two "barrel-like" modules, each formed by seven β strands. Similar to the CRH domains of gp130, the growth hormone receptor, and the erythropoietin receptor (EPO-R), these barrels are connected by a proline-rich linker that positions them at an approximately perpendicular angle (Bravo and Heath 2000). Crystallography studies of receptor/ligand complexes, epitope mapping with monoclonal antibodies, and alanine scanning mutagenesis have provided detailed insight into the composition of these complexes and the contact sites involved in ligand recognition. These data suggest that G-CSF and the G-CSF-R form a 2:2 tetrameric complex (Aritomi et al. 1999). Although it was initially proposed that this involved "pseudo-symmetric" binding of G-CSF to two sites within the CRH domain of the G-CSF-R, it now appears more likely that G-

CSF binds to one site within the CRH domain, via its type II binding motif, and to one site within the Ig domain, via type III motif binding (Layton et al. 2001). This configuration is similar to that found in the IL-6/gp130 complex (Bravo and Heath 2000).

The role of the FNIII domains in G-CSF-R function is not clear. Interestingly, the second FNIII module of the G-CSF-R was shown to confer ligand-independent activation to a chimeric G-CSF-R/gp130 receptor in Cos cells (Kurth et al. 2000). Although this suggests that the FNIII domain may be involved in the formation of an active receptor complex, the significance of this mechanism for G-CSF-R activation under more physiological conditions remains to be established.

The intracellular domain of the G-CSF-R has limited sequence homology to other hematopoietin receptor superfamily members. However, it does possess two motifs in the membrane-proximal region, called box 1 and box 2, which are also found in the EPO-R, gp130, and in the β chains of the IL-2 and IL-3 receptors (Fukunaga et al. 1991; Murakami et al. 1991). This membrane-proximal region is essential for the transduction of proliferation signals (Barge et al. 1996). The C-terminal (membrane-distal) region of the G-CSF-R contains a third conserved motif (box 3) that is shared only with gp130 (Hibi et al. 1990; Saito et al. 1992). This region has been implicated in the control of G-CSF-induced differentiation of myeloid progenitor cell lines and more recently also in the transduction of phagocytic signals in mature neutrophils (Dong et al. 1993; Fukunaga et al. 1993; Santini et al. 2003). Importantly, as will be discussed later in this review, mutations have been reported in severe congenital neutropenia (SCN) patients that result in the truncation of this C-terminal region. The cytoplasmic domain of human G-CSF-R further contains four conserved tyrosine residues, at positions 704, 729, 744, and 764 (equivalent to 703, 728, 743, and 763 in the murine G-CSF-R), which function as docking sites for multiple SH2-containing signaling proteins.

G-CSF-R expression has been demonstrated on a variety of hematopoietic cells, including myeloid progenitors, mature neutrophils, monocytes, myeloid and lymphoid leukemia cells, and normal B and T lymphocytes (Budel et al. 1989; Hanazono et al. 1990; Khwaja et al. 1993; Shimoda et al. 1993; Tsuchiya et al. 1993; Corcione et al. 1996; Morikawa et al. 1996; Matsushita and Arima 1998; Boneberg et al. 2000; Morikawa et al. 2002). G-CSF receptors have also been detected in nonhematopoietic tissues, for instance at the materno-fetal interface and on vascular endothelial cells, and in a wide variety of fetal organ tissues (McCracken et al. 1996; Calhoun et al. 1999; McCracken et al. 1999). The G-CSF-R probably plays minimal or redundant roles in embryonic development, since G-CSF-R-deficient mice are born normally, without any detectable abnormalities other than severe neutropenia (Liu et al. 1996; Hermans et al. 2002). In addition to the wt form of the G-CSF-R, at least six isoforms have been described, all of which are products of alternative mRNA splicing. The expression levels of these alternate isoforms in bone marrow progenitor cells are low or undetectable compared to the wt G-CSF-R, suggesting that their physiological role in normal myelopoiesis is minimal. However, overexpression of certain isoforms has been reported in cases of acute myeloid leukemia that result in disturbed G-CSF responses in leukemic progenitor cells (Fukunaga et al. 1990; Larsen et al. 1990; Dong et al. 1995b).

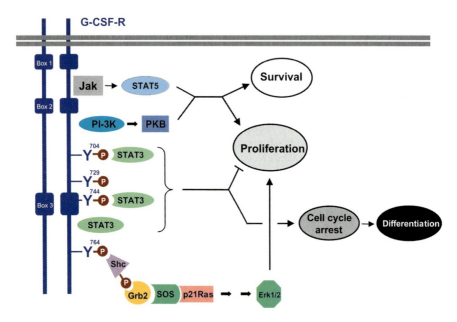

Fig. 1 Signal transduction pathways activated by the G-CSF receptor and their contribution to cellular responses of myeloid progenitor cells to G-CSF

Signaling pathways coupled to the G-CSF-R

In the past decade, the basic principles of hematopoietin receptor signaling have been elucidated. The canonical Jak/Stat pathways are generally seen as the pivotal signaling mechanisms of these receptors. Indeed, studies in knock-out models have established specific as well as more general roles for Jaks and Stats in cellular responses to growth factors and cytokines (Ihle and Kerr 1995; Ihle et al. 1995; Ihle et al. 1997). The Jak/Stat signaling components activated by G-CSF-R are Jak1, Jak2, Tyk2, Stat1, Stat3, and Stat5 (Nicholson et al. 1994; Shimoda et al. 1994; Tian et al. 1994; Tian et al. 1996; Shimoda et al. 1997).

As is the case for most other hematopoietin receptors, the p21Ras and phosphatidylinositol 3-kinase (PI-3K)/protein kinase B (PKB) signaling pathways are activated by the G-CSF-R, and both pathways were found to contribute to G-CSF-induced survival and proliferation (Fig. 1) (de Koning et al. 1998; Hunter and Avalos 1998; Ward et al. 1999b; Dong and Larner 2000; Hermans et al. 2002). Studies in the chicken B cell system DT40 suggested that activation of PI-3K depends on the presence of p55^{Lyn}. This pathway is thought to involve association of Lyn with c-Cbl, and subsequent docking of the p85 subunit of PI-3K to Y731 of Cbl (Dombrosky-Ferlan and Corey 1997; Corey et al. 1998; Grishin et al. 2000; Sinha et al. 2001).

Jak/Stat pathways

Although it has been firmly established that G-CSF activates Jak1, Jak2, and Tyk2, the specific roles of these kinases in G-CSF signaling are not clear (Nicholson et al. 1994; Shimoda et al. 1994; Shimoda et al. 1997). By employing a Jak-deficient human fibrosarcoma cell model, Shimoda and colleagues showed that Jak1, but not the other activated Jak-family members, is critical for receptor phosphorylation and Stat activation (Shimoda et al. 1997). In contrast, coexpression of dominant negative forms of either Jak1, Jak2, or Tyk2 with a wt G-CSF-R in Cos cells completely blocked G-CSF-induced Stat5 activation in these cells (Dong and Larner 2000). Moreover, Jak1-deficient mice possess normal numbers of neutrophils, which would also argue against a major and nonredundant role of Jak1 in granulopoiesis (Rodig et al. 1998). Clearly, studies in appropriate hematopoietic cell models lacking each of the Jak family members activated by the G-CSF-R are needed to resolve this issue.

Among the different Stat family members, Stat1 is only weakly and transiently activated by G-CSF and studies in Stat1-deficient mice suggest that it is redundant for granulopoiesis (de Koning et al. 1996a; Durbin et al. 1996; Meraz et al. 1996). In contrast, Stat3 is robustly activated by the G-CSF-R. Y704 and Y744 of the G-CSF-R are major docking sites for Stat3 (Fig. 1). At low ligand concentrations, Stat3 activation depends largely on the availability of at least one of these sites (de Koning et al. 1996a; Tian et al. 1996; Chakraborty et al. 1999; Ward et al. 1999b). In contrast, investigations in Ba/F3 cells, and more recently in primary bone marrow cultures, have established that at saturating G-CSF concentrations Stat3 can also be activated via a tyrosine-independent route. The latter mechanism requires the presence of the membrane-distal region of the G-CSF-R (Ward et al. 1999a; Akbarzadeh et al. 2002). Although the exact nature of this tyrosine-independent route is still unclear, this observation has led to the idea that different mechanisms for Stat3 activation might be involved in the control of steady-state granulopoiesis at low G-CSF levels (mainly tyrosine-dependent) versus "emergency" granulopoiesis initiated by increased levels of G-CSF (tyrosine-independent; Ward et al. 1999a).

The question of how Stat3 contributes to G-CSF-controlled granulopoiesis has been addressed quite extensively in both in vitro and in vivo models. Introduction of dominant negative (DN) forms of Stat3, which either prevent dimerization or DNA binding of Stat3 complexes, in myeloid cell lines resulted in a lack of growth arrest and a block in neutrophilic differentiation (Shimozaki et al. 1997; de Koning et al. 2000). Importantly, following forced G1 arrest, cells expressing DN-Stat3 fully regained their ability to differentiate, suggesting that Stat3 is required for cell cycle exit, a prerequisite for myeloid differentiation, but not for execution of the differentiation program itself (Sherr and Roberts 1995; de Koning et al. 2000). Studies in conditional knock-out mice with selective deletion of Stat3 in hematopoietic progenitor cells showed that production of functional neutrophils in vivo does not require Stat3, thereby confirming the in vitro findings that Stat3 is not essential for neutrophil differentiation per se. In fact, these conditional Stat3 knock-out mice developed a neutrophilia which was driven by a hyperproliferative response of bone marrow progenitors to G-CSF (Lee et al. 2002).

McLemore and colleagues suggested that Stat3 is not only critical for G-CSF-induced growth arrest and differentiation, but also for proliferation of myeloid progenitors, which appears partly in conflict with the data obtained in the conditional Stat3 knockout and in the cell line models (McLemore et al. 2001). They based this conclusion on a mouse model expressing a truncated G-CSF-R, in which the remaining Stat3 binding site (Y704) is mutated (d715F). The d715F mice demonstrated a complete loss of Stat3 activation in response to G-CSF and were severely neutropenic. G-CSF-driven proliferation of myeloid progenitors from d715F mice in colony cultures was almost completely restored by introduction of a constitutively active form of Stat3 (Stat3C). This suggests that Stat3 activation via Y704 plays a major role in proliferative responses. A possible explanation for the phenotypic differences between Stat3-/- and d715F mice is that in the latter model G-CSF signaling is aberrant in more ways than in just its inability to activate Stat3. For instance, internalization of the truncated receptors is severely hampered and signaling abilities and signal duration are quite drastically altered compared to the wt G-CSF-R (see below; Hermans et al. 1999). Additionally, the constitutively active Stat3 protein is an oncoprotein that may perturb multiple signaling mechanisms and thus synergize with G-CSF in evoking proliferative responses (Bromberg et al. 1999). The combination of the truncated G-CSF-R with the constitutively active Stat3 in the study by McLemore et al. might therefore overestimate of the role of Stat3 in normal granulopoiesis.

The mechanisms by which Stat3 contributes to cell cycle exit in myeloid progenitor cells remain unclear. The cyclin-dependent kinase (cdk) inhibitor $p27^{Kip1}$ has been proposed to play a role in this process (de Koning et al. 2000). G-CSF induces expression of $p27^{Kip1}$ in 32D cells. Dominant-negative forms of Stat3 completely block this G-CSF-induced $p27^{Kip1}$ expression. Furthermore, a putative Stat3 binding site was identified in the promoter region of $p27^{Kip1}$ that was functional in both electrophoretic mobility shift assays and in luciferase reporter assays. Finally, myeloid progenitors from $p27^{Kip1}$-deficient mice showed significantly increased proliferation and reduced differentiation in response to G-CSF, compared with wt controls. Taken together, these findings suggested that Stat3 controls cell cycle arrest of myeloid cells, at least partly, via transcriptional upregulation of $p27^{Kip1}$. It is important to note, however, that transcription of $p27^{Kip1}$ is also, and arguably more robustly, induced by transcription factors of the Forkhead family, which are negatively controlled by phosphorylation through the PI-3K/PKB pathway (Kops et al. 1999; Medema et al. 2000). Interestingly, recent studies in HepG2 cells indicate that one of these factors, FKHR, acts as a coactivator of Stat3 in IL-6 induced transcriptional activity (Kortylewski et al. 2003). Whether this also applies to G-CSF-induced upregulation of $p27^{Kip1}$ remains to be addressed.

Stat5 has been implicated in proliferation and survival signals provided by the G-CSF-R (Dong et al. 1998). The role of Stat5 in steady-state granulopoiesis appears limited, as double-knockout mice lacking both the Stat5A and Stat5B isoforms have only moderately reduced numbers of CFU-G and no overt neutropenia (Teglund et al. 1998). Whether Stat5 is involved in G-CSF-driven emergency granulopoiesis has not been established. Irrespective of its role in nonmalignant granulopoiesis, Stat5 may be a crucial player in the pathogenesis of myeloid malignancies. For instance, the transforming abilities of the Tel-Jak2 fusion protein, a hallmark of a specific subset of myeloid leukemia, depend entirely on the presence of Stat5 (Schwaller et al. 2000).

p21Ras/MAPKinase pathways

Y764 of the G-CSF-R plays a major role in proliferation signaling in cell line models as well as in primary myeloid progenitor cells (de Koning et al. 1998; Akbarzadeh et al. 2002; Hermans et al. 2002). Once phosphorylated, Y764 forms a binding site for the SH2 domains of Shc, Grb2, and SHP-2, signaling intermediates of the p21Ras pathway (de Koning et al. 1996b; Rausch and Marshall 1997; de Koning et al. 1998; Ward et al. 1999b). Grb2 can also be recruited via docking to Shc and SHP-2 (Bennett et al. 1994; van der Geer et al. 1996; Vogel and Ullrich 1996; Harmer and DeFranco 1997). Loss of Y764 results in a significant reduction of p21Ras activation, and accelerated neutrophil differentiation (Bashey et al. 1994; Rausch and Marshall 1997; de Koning et al. 1998). Interestingly, SCN-derived G-CSF-R truncation mutants that lack the receptor C-terminus gain the ability of p21Ras activation by an alternative mechanism, probably involving the SHP-2/Grb2 route linked to Y704 of the G-CSF-R (de Koning et al. 1998; Ward et al. 1999b).

Studies utilizing antisense technology, specific pharmacological inhibitors and dominant negative forms of signaling intermediates identified the Raf-Mek-Erk MAPkinase cascade as the major effector pathway downstream from p21Ras responsible for proliferative signaling in cell lines as well as in primary myeloid progenitor cells (Fig. 1) (Bashey et al. 1994; Muszynski et al. 1995; Keller et al. 1996; Darley and Burnett 1999; Rausch and Marshall 1999; Baumann et al. 2001; Akbarzadeh et al. 2002; Hermans et al. 2002; Koay et al. 2002). Activation of other MAPKs downstream of p21Ras, i.e., the p38MAPK and Jun N-terminal kinase (JNK) is also controlled mainly via Y764, but the role of these kinases in G-CSF signaling is still unclear (Rausch and Marshall 1997; Rausch and Marshall 1999).

Negative regulation of G-CSF signaling

The inhibition of cytokine responses is governed by multiple mechanisms, including dephosphorylation of signaling molecules by phosphatases, receptor endocytosis, and proteasomal targeting. Mechanisms that have been implicated in the downregulation G-CSF signaling are discussed below.

SHP-1

The role of the SH2 domain-containing protein tyrosine phosphatase SHP-1 as a negative regulator of granulopoiesis has been established utilizing so-called moth-eaten (me^v) mice (Yi et al. 1992; David et al. 1995), which possess a mutation in the SHP-1 gene resulting in reduced phosphatase activity (Tsui et al. 1993). These mice exhibit aberrant regulation in several myeloid and lymphoid lineages, including substantial increases in the number of immature granulocytes (Kozlowski et al. 1993; Shultz et al. 1993; Tapley et al. 1997). SHP-1 protein levels are increased in a posttranscriptional manner during G-CSF-induced differentiation of 32D cells. Ectopic overexpression of SHP-1 in these cells inhibited proliferation and stimulated differentiation, whereas introduction of a phosphatase-dead SHP-1 mutant gave the opposite result (Ward et al. 2000b). In contrast to the EPO-R or the GM-CSF/IL-3/IL-5-R common β chains, G-CSF-R tyrosines do not serve as docking sites for the SH2 domain of SHP-1, suggesting that intermediate signaling molecules may be involved in the recruitment of SHP-1 into the G-CSF-R complex (Tapley et al. 1997; Ward et al. 2000b; Dong et al. 2001).

SHIP

A 145-kD phosphorylated protein was detected following G-CSF stimulation in both Shc and in Grb2 immunoprecipitations. The formation of these complexes depended on the presence of Y764 of the G-CSF-R (de Koning et al. 1996b). This protein was later identified as the SH2-containing inositol phosphatase (SHIP) protein (Hunter and Avalos 1998). Studies in SHIP-deficient mice showed that this phosphatase is important for modulating hematopoietic signaling, particularly in the myeloid lineage. SHIP-/- mice die early, most likely due to the extensive infiltration of myeloid cells observed in the lungs. The numbers of neutrophils and monocytes in these mice are increased, which is due to elevated numbers of myeloid progenitors in the bone marrow (Helgason et al. 1998). Furthermore, survival of neutrophils lacking SHIP is prolonged following apoptosis-inducing stimuli or growth factor withdrawal. Finally, PI(3,4,5)P3 accumulation and PKB activation are both increased and prolonged in SHIP-/- cells. Taken together, these data suggest a role for SHIP as a negative regulator of growth factor-mediated PI-3K/PKB activation and survival of myeloid cells (Liu et al. 1999).

SOCS proteins

Suppressor of cytokine signaling (SOCS) proteins downregulate cytokine responses by competing with positively acting signaling substrates for receptor tyrosine docking, by inhibiting the activity of receptor-associated kinases, and by targeting signaling molecules for proteasomal degradation (Matsumoto et al. 1997; Zhang et al. 1999). For example, SOCS1 binds to Jak kinases via its SH2 domain and can inhibit kinase activity directly (Masuhara et al. 1997; Naka et al. 1997; Nicholson et al. 1999; Yasukawa et al. 1999). Other members of the SOCS protein family, such as SOCS3, require recruitment to phosphotyrosines in activated receptors for signal inhibition (Cohney et al. 1999; Eyckerman et al. 2000; Nicholson et al. 2000; Schmitz et al. 2000).

Expression of SOCS proteins is under the direct transcriptional control of Stats. It is thus conceivable that SOCS proteins are also involved in downmodulation of G-CSF signaling as the direct consequence of the robust and sustained activation of Stat3 (Naka et al. 1997; Auernhammer et al. 1999; Davey et al. 1999; Yasukawa et al. 2000). Among the different SOCS family members upregulated by G-CSF, SOCS3 is most prominently induced (Starr et al. 1997; Hortner et al. 2002). G-CSF-R Y729 forms the major recruitment site for SOCS3 (Hermans et al. 2002; Hortner et al. 2002). Colony cultures of bone marrow cells transduced with tyrosine add-back and substitution mutants of the G-CSF-R supported the functional significance of this negative feedback mechanism involving Y729 (Akbarzadeh et al. 2002; Hermans et al. 2002). Significantly, the SOCS3 recruitment site Y729 is lost in truncated SCN/AML-derived forms of G-CSF-R. In addition, these mutant G-CSF-R are hampered in their ability to induce SOCS3 transcription (G.J.M. van de Geijn et al., in prep.).

G-CSF-induced SOCS3 expression is severely reduced in Stat3-/- mice, raising the possibility that SOCS3 is the major Stat3 target responsible for inhibition of G-CSF signaling (Lee et al. 2002). Although attractive, this hypothesis does not seem to apply to steady state granulopoiesis, because SOCS3-deficient mice do not present with the neutrophilia observed in conditional Stat3-deficient mice (Marine et al. 1999; Roberts et al. 2001;

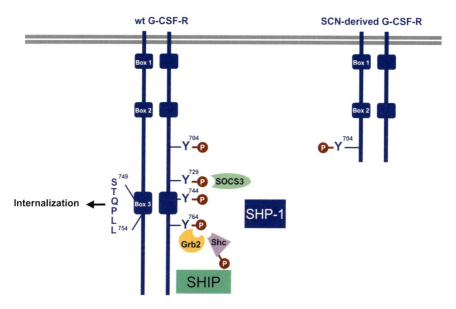

Fig. 2 Activation of multiple negative feedback mechanisms is lost in SCN-derived truncated G-CSF-R mutants. The wt G-CSF-R activates the phosphatases SHIP and SHP-1 and recruits SOCS3. The C-terminus (>amino acid 715) is required for activation of SHP-1, but the exact mechanism involved is not known. The truncated, SCN-derived G-CSF-R lacks the domains required for activation of SHIP, SHP-1, the major SOCS3 recruitment site Y729, and a dileucine-based internalization motif. Loss of all these mechanisms probably contributes in an additive way to the dominant hyperproliferative function of this mutant receptor

Takahashi et al. 2003). Perhaps SOCS3-mediated inhibition only becomes efficient during G-CSF-induced "emergency" granulopoiesis.

Receptor endocytosis

Following ligand-binding, growth factor receptors are usually incorporated into clathrin-coated pits, internalized, and subsequently either recycled back to the plasma membrane, retained in the endosomal compartment, or targeted for lysosomal degradation (Ceresa and Schmid 2000). Truncated G-CSF-Rs are severely impaired in internalization (Hermans et al. 1999; Hunter and Avalos 1999; Ward et al. 1999d). This is in part due to the loss of a serine-type dileucine motif in box 3 (amino acids 749–755) and the immediate downstream sequence stretch of amino acids 756–769 (Fig. 2; Ward et al. 1999d; L.H.J. Aarts et al., in prep.). Mutation of this dileucine motif reduced receptor endocytosis and delayed the attenuation of signaling, as well as the onset of G-CSF-induced differentiation (Ward et al. 1999d; L.H.J. Aarts et al., in prep.). Interestingly, increasing evidence suggests that receptor internalization and intracellular trafficking does not only serve as an inhibitory mechanism but may also be required for appropriate spatio-temporal activation of the full complement of signal transduction proteins (Ceresa and Schmid 2000; Di Fiore and De Camilli 2001; McPherson et al. 2001; Miller and Lefkowitz 2001). It is thus conceivable that some of the specific responses of myeloid cells to G-CSF rely on G-CSF-R internalization and intracellular trafficking.

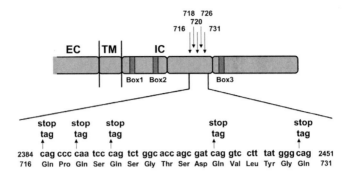

Fig. 3 Nonsense mutations resulting in truncation of the G-CSF-R C-terminus in SCN patients. Mutations are confined to a region between amino acids 716 and 731. *EC* extracellular part, *TM* transmembrane region, *IC* intracellular part of the G-CSF-R

G-CSF-R defects in myeloid disorders

A number of mutations or rare polymorphisms in the *GCSFR* gene have been reported in myeloid disorders, and these were found to perturb signaling functions of the receptor. Mutations are found most frequently in SCN (but not in cyclic or idiopathic neutropenia) and rarely in myelodysplasia (MDS) and de novo AML. Elucidation of the functional consequences of these abnormalities has contributed to our understanding of the role of specific domains of the G-CSF-R in signaling.

Nonsense mutations in a critical glutamine-rich stretch, which result in C-terminal truncation of the G-CSF-R, are the most frequent mutations found in SCN (Fig. 3). Clones harboring such acquired mutations are detected in the neutropenic phase of the disease in approximately 20% of patients (Dong et al. 1995a; Dong et al. 1997). In some cases, affected myeloid cells arise from minority clones, originally making up only 1%–2% of the myeloid progenitor cell compartment. However, clones with G-CSF-R mutations become overt in more than 80% of the SCN cases upon progression to MDS and AML, suggesting that G-CSF-R truncations represent a critical step in the expansion of the (pre-) leukemic clones (Freedman and Alter 2002). An important question in this context is how G-CSF treatment contributes to the outgrowth of the leukemia. In a recent update from the Severe Chronic Neutropenia International Registry, evolution of SCN to MDS or AML was reported in 35 of 387 patients with congenital neutropenia with a cumulative risk of 13% after eight years of G-CSF treatment, but there was no apparent relationship to duration or dose of G-CSF treatment (Dale et al. 2003).

The role of these truncation mutations in leukemic transformation has been analyzed in further detail in mouse models in which the nonsense mutation was introduced in the G-CSF-R gene by knock-in strategies (Hermans et al. 1998; McLemore et al. 1998). Although insufficient to cause leukemia themselves, these mutations were recently found to cooperate with additional oncogenic hits, such as loss of the DNA repair protein MSH2, to accelerate tumorigenesis. Interestingly, preliminary results also suggest an association between loss of MSH2 function and G-CSF-R mutations in SCN patients at high risk for AML/MDS progression, supporting this murine data (J.M. Prasher et al., in prep.). Mice expressing the truncated G-CSF-R exhibit hyperproliferation of myeloid progenitor cells in response to G-CSF (Hermans et al. 1998; McLemore et al. 1998). Multiple signaling abnormalities have been linked with this hyperproliferation, including defective receptor internalization, increased and sustained activation of Stat5 and PI-3K/PKB, loss of negative feedback by SHP-1, and loss of specific docking sites for the negative regulators SOCS3 and SHIP (Fig. 2; Hunter and Avalos 1998; Hermans et al. 1999; Hunter and Avalos 1999; Ward et al. 1999d; Dong and Larner 2000; Hortner et al. 2002).

A second type of G-CSF-R mutation was found in an SCN patient who failed to respond to G-CSF treatment (Ward et al. 1999c). This mutation, located in the extracellular domain, changes a conserved proline residue in the "hinge" motif located between the NH2- and COOH-terminal barrels of the CRH domain, which was proposed to prevent the formation of 2:2 ligand/receptor complexes. Contrary to the C-terminal truncations, this mutant receptor showed drastically reduced activation of Stat5 and was severely hampered in proliferation and cell survival signaling in 32D cells, while differentiation-inducing properties were retained.

In MDS without a history of SCN, specific mutations in the GCSFR gene have thus far not been reported. On the other hand, Awaya et al. found an increased occurrence of a novel splice variant of G-CSF-R with an alteration in the juxtamembrane region of the receptor (Awaya et al. 2002). Via an as yet unknown mechanism, this variant conferred increased proliferative signals in response to G-CSF compared to the wt G-CSF-R. However, because this receptor variant is also found at low frequencies (2%) in normal bone marrow cells and is still only detectable in less than 8% of the myeloid progenitor cells in MDS, its role in the pathogenesis of MDS remains uncertain.

In de novo AML, activating mutations in receptor tyrosine kinases FTL3 and c-kit occur in more than 25% of cases and have a significant impact on disease prognosis (Longley et al. 2001; Gilliland 2002; Kiyoi and Naoe 2002; Schnittger et al. 2002; Moreno et al. 2003). In contrast, mutations in hematopoietin receptors, including G-CSF-R, have only very rarely been detected. A mutation leading to an overexpression of a nonfunctional splice variant of G-CSF-R was reported in 1 out of 70 cases analyzed (Dong et al. 1995b). This variant receptor has the alternative 34 C-terminal amino acids of the class IV G-CSF-R (alternatively known as D-7), linked to amino acid 682, which is just C-terminal of box-2. It thus lacks most of the functional domains, including all the tyrosine-based docking motifs, which explains why it lacks most of its signaling abilities. Although this case so far appears to be unique, altered ratios of Class I(wt)/ClassIV G-CSF-R levels have been reported in more than 50% of AML samples, which could be suggestive of a more general role for abnormal G-CSF-R function in AML (White et al. 1998). Significantly, even at relatively low levels of expression, the Class IV variant was reported to interfere with differentiation induction mediated via the wt G-CSF-R in 32Dcl3 cells (White et al. 2000).

More recently, an activating mutation in the transmembrane domain of G-CSF-R was reported in 2/555 AML patients (Forbes et al. 2002). This mutation conferred growth factor independence on Ba/F3 cells and results in the constitutive phosphorylation of signaling substrates (Jak2, Stat3, ERK1, ERK2) as well as the receptor itself. The observation that point mutations in the TM domain can lead to constitutive receptor activation corroborates earlier observations in experimental leukemia models expressing constitutively active forms of the β common chain of IL-3/IL-5/GM-CSF receptor or c-MPL as a result of mutations in the TM domain (Jenkins et al. 1995; Onishi et al. 1996; Jenkins et al. 1998). These results demonstrate that the TM domain of the G-CSF-R (and other hematopoietin receptors) does not simply form a membrane anchor, but also contributes to the conformation for receptor complexes, thereby influencing their signaling properties.

Concluding remarks

Ample evidence obtained from both in vivo and in vitro models has established that G-CSF and its receptor fulfill nonredundant functions in the regulation of neutrophil production. However, the fact that G-CSF and G-CSF-R deficient mice still contain mature neutrophils, albeit at strongly reduced levels, implies that neutrophilic differentiation does not depend on signals emanating from the G-CSF-R. Indeed, the signaling pathways discussed in this review are involved on the control of cell proliferation, cell survival, or induction of a G1 arrest, but none of them appear to be required for the execution of the differentiation process itself. It has been well established that the combinatorial action of transcription factors determines hematopoietic cell commitment and differentiation into the various hematopoietic lineages (Tenen et al. 1997; Ward et al. 2000a; Tenen 2001; Friedman 2002). Interestingly, recent studies have indicated that transcription factors implicated in the control of myeloid differentiation depend on signaling pathways for their expression levels and/or their full spectrum of activities. For instance, this applies to C/EBPα and C/EBPε, members of the CCAAT/enhancer binding protein family that are essential for early and later stages of neutrophilic differentiation, respectively (Friedman 2002). Specifically, the p21Ras pathway enhances C/EBPα activity by phosphorylation of a serine residue (Ser 248) and mutation of this residue inhibited the ability of C/EBPα to induce granulocytic differentiation (Behre et al. 2002). In the case of C/EBPε, a direct stimulatory role for the G-CSF-R in the induction of C/EBPε expression was reported, which was found to contribute to neutrophilic differentiation (Nakajima and Ihle 2001). These observations fit into a scenario in which the G-CSF-R provides signals that act in concert with transcription factors to tightly control neutrophilic differentiation under both steady state and emergency conditions. Further studies specifically directed towards this interplay between G-CSF-induced signaling events and transcriptional control of myelopoiesis might help to understand the complex pathogenesis of hematological diseases characterized by a block in myeloid differentiation.

Acknowledgements. This work is supported by the Dutch Cancer Society "Koningin Wilhelmina Fonds."

References

Akbarzadeh S, Ward AC, McPhee DO, Alexander WS, Lieschke GJ, Layton JE (2002) Tyrosine residues of the granulocyte colony-stimulating factor receptor transmit proliferation and differentiation signals in murine bone marrow cells. Blood 99:879–887

Aritomi M, Kunishima N, Okamoto T, Kuroki R, Ota Y, Morikawa K (1999) Atomic structure of the GCSF-receptor complex showing a new cytokine-receptor recognition scheme. Nature 401:713–717

Auernhammer CJ, Bousquet C, Melmed S (1999) Autoregulation of pituitary corticotroph SOCS-3 expression: characterization of the murine SOCS-3 promoter. Proc Natl Acad Sci USA 96:6964–6969

Avalos BR, Gasson JC, Hedvat C, Quan SG, Baldwin GC, Weisbart RH, Williams RE, Golde DW, DiPersio JF (1990) Human granulocyte colony-stimulating factor: biologic activities and receptor characterization on hematopoietic cells and small cell lung cancer cell lines. Blood 75:851–857

Awaya N, Uchida H, Miyakawa Y, Kinjo K, Matsushita H, Nakajima H, Ikeda Y, Kizaki M (2002) Novel variant isoform of G-CSF receptor involved in induction of proliferation of FDCP-2 cells: relevance to the pathogenesis of myelodysplastic syndrome. J Cell Physiol 191:327–335

Barge RM, de Koning JP, Pouwels K, Dong F, Lowenberg B, Touw IP (1996) Tryptophan 650 of human granulocyte colony-stimulating factor (G-CSF) receptor, implicated in the activation of JAK2, is also required for G-CSF-mediated activation of signaling complexes of the p21ras route. Blood 87:2148–2153

Bashey A, Healy L, Marshall CJ (1994) Proliferative but not nonproliferative responses to granulocyte colony-stimulating factor are associated with rapid activation of the p21ras/MAP kinase signaling pathway. Blood 83:949–957

Baumann MA, Paul CC, Lemley-Gillespie S, Oyster M, Gomez-Cambronero J (2001) Modulation of MEK activity during G-CSF signaling alters proliferative versus differentiative balancing. Am J Hematol 68:99–105

Bazan JF (1990) Structural design and molecular evolution of a cytokine receptor superfamily. Proc Natl Acad Sci USA 87:6934–6938

Behre G, Singh SM, Liu H, Bortolin LT, Christopeit M, Radomska HS, Rangatia J, Hiddemann W, Friedman AD, Tenen DG (2002) Ras signaling enhances the activity of C/EBP α to induce granulocytic differentiation by phosphorylation of serine 248. J Biol Chem 277:26293–26299

Bennett A, Tang T, Sugimoto S, Walsh C, Neel B (1994) Protein-tyrosine-phosphatase SHPTP2 couples platelet-derived growth factor receptor $\{\beta\}$ to Ras. Proc Natl Acad Sci USA 91:7335–7339

Berliner N, Hsing A, Graubert T, Sigurdsson F, Zain M, Bruno E, Hoffman R (1995) Granulocyte colony-stimulating factor induction of normal human bone marrow progenitors results in neutrophil-specific gene expression. Blood 85:799–803

Boneberg EM, Hareng L, Gantner F, Wendel A, Hartung T (2000) Human monocytes express functional receptors for granulocyte colony-stimulating factor that mediate suppression of monokines and interferon-γ. Blood 95:270–276

Bravo J, Heath JK (2000) New EMBO members' review: receptor recognition by gp130 cytokines. EMBO J. 19:2399–2411

Bromberg JF, Wrzeszczynska MH, Devgan G, Zhao Y, Pestell RG, Albanese C, Darnell JE, Jr. (1999) Stat3 as an oncogene. Cell 98:295–303

Budel LM, Touw IP, Delwel R, Lowenberg B (1989) Granulocyte colony-stimulating factor receptors in human acute myelocytic leukemia. Blood 74:2668–2673

Calhoun DA, Donnelly WH, Jr., Du Y, Dame JB, Li Y, Christensen RD (1999) Distribution of granulocyte colony-stimulating factor (G-CSF) and G-CSF-receptor mRNA and protein in the human fetus. Pediatr Res 46:333–338

Ceresa BP, Schmid SL (2000) Regulation of signal transduction by endocytosis. Curr Opin Cell Biol 12:204–210

Chakraborty A, Dyer KF, Cascio M, Mietzner TA, Tweardy DJ (1999) Identification of a novel Stat3 recruitment and activation motif within the granulocyte colony-stimulating factor receptor. Blood 93:15–24

Cohney SJ, Sanden D, Cacalano NA, Yoshimura A, Mui A, Migone TS, Johnston JA (1999) SOCS-3 is tyrosine phosphorylated in response to interleukin-2 and suppresses STAT5 phosphorylation and lymphocyte proliferation. Mol Cell Biol 19:4980–4988

Corcione A, Corrias MV, Daniele S, Zupo S, Spriano M, Pistoia V (1996) Expression of granulocyte colony-stimulating factor and granulocyte colony-stimulating factor receptor genes in partially overlapping monoclonal B-cell populations from chronic lymphocytic leukemia patients. Blood 87:2861–2869

Corey SJ, Dombrosky-Ferlan PM, Zuo S, Krohn E, Donnenberg AD, Zorich P, Romero G, Takata M, Kurosaki T (1998) Requirement of Src kinase Lyn for induction of DNA synthesis by granulocyte colony-stimulating factor. J Biol Chem 273:3230–3235

Cosman D (1993) The hematopoietin receptor superfamily. Cytokine 5:95–106

Dale DC, Bonilla MA, Davis MW, Nakanishi AM, Hammond WP, Kurtzberg J, Wang W, Jakubowski A, Winton E, Lalezari P, et al. (1993) A randomized controlled phase III trial of recombinant human granulocyte colony-stimulating factor (filgrastim) for treatment of severe chronic neutropenia. Blood 81:2496–2502

Dale DC, Cottle TE, Fier CJ, Bolyard AA, Bonilla MA, Boxer LA, Cham B, Freedman MH, Kannourakis G, Kinsey SE, Davis R, Scarlata D, Schwinzer B, Zeidler C, Welte K (2003) Severe chronic neutropenia: treatment and follow-up of patients in the Severe Chronic Neutropenia International Registry. Am J Hematol 72:82–93

D'Andrea AD, Lodish HF, Wong GG (1989) Expression cloning of the murine erythropoietin receptor. Cell 57:277–285

Darley RL, Burnett AK (1999) Mutant RAS inhibits neutrophil but not macrophage differentiation and allows continued growth of neutrophil precursors. Exp Hematol 27:1599–1608

Davey HW, McLachlan MJ, Wilkins RJ, Hilton DJ, Adams TE (1999) STAT5b mediates the GH-induced expression of SOCS-2 and SOCS-3 mRNA in the liver. Mol Cell Endocrinol 158:111–116

David M, Chen H, Goelz S, Larner A, Neel B (1995) Differential regulation of the α/βinterferon-stimulated Jak/Stat pathway by the SH2 domain-containing tyrosine phosphatase SHPTP1. Mol Cell Biol 15:7050–7058 de Koning JP, Dong F, Smith L, Schelen AM, Barge RM, van der Plas DC, Hoefsloot LH, Lowenberg B, Touw IP (1996a) The membrane-distal cytoplasmic region of human granulocyte colony-stimulating factor receptor is required for STAT3 but not STAT1 homodimer formation. Blood 87:1335–1342

de Koning JP, Schelen AM, Dong F, van Buitenen C, Burgering BM, Bos JL, Lowenberg B, Touw IP (1996b) Specific involvement of tyrosine 764 of human granulocyte colony-stimulating factor receptor in signal transduction mediated by p145/Shc/GRB2 or p90/GRB2 complexes. Blood 87:132–140

de Koning JP, Soede-Bobok AA, Schelen AM, Smith L, van Leeuwen D, Santini V, Burgering BM, Bos JL, Lowenberg B, Touw IP (1998) Proliferation signaling and activation of Shc, p21Ras, and Myc via tyrosine 764 of human granulocyte colony-stimulating factor receptor. Blood 91:1924–1933

de Koning JP, Soede-Bobok AA, Ward AC, Schelen AM, Antonissen C, van Leeuwen D, Lowenberg B, Touw IP (2000) STAT3-mediated differentiation and survival and of myeloid cells in response to granulocyte colony-stimulating factor: role for the cyclin-dependent kinase inhibitor p27(Kip1). Oncogene 19:3290–3298

Demetri GD, Griffin JD (1991) Granulocyte colony-stimulating factor and its receptor. [Review] [170 refs]. Blood 78:2791–2808

Di Fiore PP, De Camilli P (2001) Endocytosis and signaling: an inseparable partnership. Cell 106:1–4

Dombrosky-Ferlan PM, Corey SJ (1997) Yeast two-hybrid in vivo association of the Src kinase Lyn with the proto-oncogene product Cbl but not with the p85 subunit of PI 3-kinase. Oncogene 14:2019–2024

Dong F, Larner AC (2000) Activation of Akt kinase by granulocyte colony-stimulating factor (G-CSF): evidence for the role of a tyrosine kinase activity distinct from the Janus kinases. Blood 95:1656–1662

Dong F, van Buitenen C, Pouwels K, Hoefsloot LH, Lowenberg B, Touw IP (1993) Distinct cytoplasmic regions of the human granulocyte colony-stimulating factor receptor involved in induction of proliferation and maturation. Mol Cell Biol 13:7774–7781

Dong F, Brynes RK, Tidow N, Welte K, Lowenberg B, Touw IP (1995a) Mutations in the gene for the granulocyte colony-stimulating-factor receptor in patients with acute myeloid leukemia preceded by severe congenital neutropenia [see comments]. N Engl J Med 333:487–493

Dong F, van Paassen M, van Buitenen C, Hoefsloot LH, Lowenberg B, Touw IP (1995b) A point mutation in the granulocyte colony-stimulating factor receptor (G-CSF-R) gene in a case of acute myeloid leukemia results in the overexpression of a novel G-CSF-R isoform. Blood 85:902–911

Dong F, Dale DC, Bonilla MA, Freedman M, Fasth A, Neijens HJ, Palmblad J, Briars GL, Carlsson G, Veerman AJ, Welte K, Lowenberg B, Touw IP (1997) Mutations in the granulocyte colony-stimulating factor receptor gene in patients with severe congenital neutropenia. Leukemia 11:120–125

Dong F, Liu X, de Koning JP, Touw IP, Henninghausen L, Larner A, Grimley PM (1998) Stimulation of Stat5 by granulocyte colony-stimulating factor (G-CSF) is modulated by two distinct cytoplasmic regions of the G-CSF receptor. J Immunol 161:6503–6509

Dong F, Qiu Y, Yi T, Touw IP, Larner AC (2001) The carboxyl terminus of the granulocyte colony-stimulating factor receptor, truncated in patients with severe congenital neutropenia/acute myeloid leukemia, is required for SH2-containing phosphatase-1 suppression of Stat activation. J Immunol 167:6447–6452

Durbin JE, Hackenmiller R, Simon MC, Levy DE (1996) Targeted disruption of the mouse Stat1 gene results in compromised innate immunity to viral disease. Cell 84:443–450

Eyckerman S, Broekaert D, Verhee A, Vandekerckhove J, Tavernier J (2000) Identification of the Y985 and Y1077 motifs as SOCS3 recruitment sites in the murine leptin receptor. FEBS Lett 486:33–37

Forbes LV, Gale RE, Pizzey A, Pouwels K, Nathwani A, Linch DC (2002) An activating mutation in the transmembrane domain of the granulocyte colony-stimulating factor receptor in patients with acute myeloid leukemia. Oncogene 21:5981–5989

Freedman MH, Alter BP (2002) Risk of myelodysplastic syndrome and acute myeloid leukemia in congenital neutropenias. Semin Hematol 39:128–133

Friedman AD (2002) Transcriptional regulation of granulocyte and monocyte development. Oncogene 21:3377–3390

Fukunaga R, Seto Y, Mizushima S, Nagata S (1990) Three different mRNAs encoding human granulocyte colony-stimulating factor receptor. Proc Natl Acad Sci USA 87:8702–8706

Fukunaga R, Ishizaka-Ikeda E, Pan CX, Seto Y, Nagata S (1991) Functional domains of the granulocyte colony-stimulating factor receptor. EMBO J 10:2855–2865

Fukunaga R, Ishizaka-Ikeda E, Nagata S (1993) Growth and differentiation signals mediated by different regions in the cytoplasmic domain of granulocyte colony-stimulating factor receptor. Cell 74:1079–1087

Gilliland DG (2002) Molecular genetics of human leukemias: New insights into therapy. Semin Hematol 39:6–11

Grishin A, Sinha S, Roginskaya V, Boyer MJ, Gomez-Cambronero J, Zuo S, Kurosaki T, Romero G, Corey SJ (2000) Involvement of Shc and Cbl-PI 3-kinase in Lyn-dependent proliferative signaling pathways for G-CSF. Oncogene 19:97–105

Hanazono Y, Hosoi T, Kuwaki T, Matsuki S, Miyazono K, Miyagawa K, Takaku F (1990) Structural analysis of the receptors for granulocyte colony-stimulating factor on neutrophils. Exp Hematol 18:1097–1103

Harmer SL, DeFranco AL (1997) Shc contains two Grb2 binding sites needed for efficient formation of complexes with SOS in B lymphocytes. Mol Cell Biol 17:4087–4095

Helgason CD, Damen JE, Rosten P, Grewal R, Sorensen P, Chappel SM, Borowski A, Jirik F, Krystal G, Humphries RK (1998) Targeted disruption of SHIP leads to hemopoietic perturbations, lung pathology, and a shortened life span. Genes Dev 12:1610–1620

Hermans MH, Ward AC, Antonissen C, Karis A, Lowenberg B, Touw IP (1998) Perturbed granulopoiesis in mice with a targeted mutation in the granulocyte colony-stimulating factor receptor gene associated with severe chronic neutropenia. Blood 92:32–39

Hermans MH, Antonissen C, Ward AC, Mayen AE, Ploemacher RE, Touw IP (1999) Sustained receptor activation and hyperproliferation in response to granulocyte colony-stimulating factor (G-CSF) in mice with a severe congenital neutropenia/acute myeloid leukemia-derived mutation in the G-CSF receptor gene. J Exp Med 189:683–692

Hermans MH, Van De Geijn GJ, Antonissen C, Gits J, Van Leeuwen D, Ward AC, Touw IP (2003) Signaling mechanisms coupled to tyrosines in the granulocyte colony-stimulating factor receptor orchestrate G-CSF-induced expansion of myeloid progenitor cells. Blood 101:2584–2590

Hibi M, Murakami M, Saito M, Hirano T, Taga T, Kishimoto T (1990) Molecular cloning and expression of an IL-6 signal transducer, gp130. Cell 63:1149–1157

Hortner M, Nielsch U, Mayr LM, Johnston JA, Heinrich PC, Haan S (2002) Suppressor of cytokine signaling-3 is recruited to the activated granulocyte-colony stimulating factor receptor and modulates its signal transduction. J Immunol 169:1219–1227

Hunter MG, Avalos BR (1998) Phosphatidylinositol 3'-kinase and SH2-containing inositol phosphatase (SHIP) are recruited by distinct positive and negative growth-regulatory domains in the granulocyte colony-stimulating factor receptor. J Immunol 160:4979–4987

Hunter MG, Avalos BR (1999) Deletion of a critical internalization domain in the G-CSFR in acute myelogenous leukemia preceded by severe congenital neutropenia. Blood 93:440–446

Ihle JN, Kerr IM (1995) Jaks and Stats in signaling by the cytokine receptor superfamily. Trends Genet 11:69–74

Ihle JN, Witthuhn BA, Quelle FW, Yamamoto K, Silvennoinen O (1995) Signaling through the hematopoietic cytokine receptors. Annu Rev Immunol 13:369–398

Ihle JN, Nosaka T, Thierfelder W, Quelle FW, Shimoda K (1997) Jaks and Stats in cytokine signaling. Stem Cells 1:105–111

Jenkins BJ, D'Andrea R, Gonda TJ (1995) Activating point mutations in the common β subunit of the human GM-CSF, IL-3, and IL-5 receptors suggest the involvement of β subunit dimerization and cell type-specific molecules in signaling. EMBO J 14:4276–4287

Jenkins BJ, Blake TJ, Gonda TJ (1998) Saturation mutagenesis of the β subunit of the human granulocyte-macrophage colony-stimulating factor receptor shows clustering of constitutive mutations, activation of ERK MAP kinase and STAT pathways, and differential β subunit tyrosine phosphorylation. Blood 92:1989–2002

Keller JR, Ruscetti FW, Heidecker G, Linnekin DM, Rapp U, Troppmair J, Gooya J, Muszynski KW (1996) The effect of c-raf antisense oligonucleotides on growth factor-induced proliferation of hematopoietic cells. Curr Top Microbiol Immunol 211:43–53

Khwaja A, Carver J, Jones HM, Paterson D, Linch DC (1993) Expression and dynamic modulation of the human granulocyte colony-stimulating factor receptor in immature and differentiated myeloid cells. Br J Haematol 85:254–259

Kiyoi H, Naoe T (2002) FLT3 in human hematologic malignancies. Leuk Lymphoma 43:1541–1547

Koay DC, Nguyen T, Sartorelli AC (2002) Distinct region of the granulocyte colony-stimulating factor receptor mediates proliferative signaling through activation of Janus kinase 2 and p44/42 mitogen-activated protein kinase. Cell Signal 14:239–247

Kops GJ, de Ruiter ND, De Vries-Smits AM, Powell DR, Bos JL, Burgering BM (1999) Direct control of the Forkhead transcription factor AFX by protein kinase B. Nature 398:630–634

Kortylewski M, Feld F, Kruger K-D, Bahrenberg G, Roth RA, Joost H-G, Heinrich PC, Behrmann I, Barthel A (2003) Akt Modulates STAT3-mediated gene expression through a FKHR (FOXO1a)-dependent Mechanism. J Biol Chem 278:5242–5249

Kozlowski M, Mlinaric-Rascan I, Feng GS, Shen R, Pawson T, Siminovitch KA (1993) Expression and catalytic activity of the tyrosine phosphatase PTP1C is severely impaired in motheaten and viable motheaten mice. J Exp Med 178:2157–2163

Kurth I, Horsten U, Pflanz S, Timmermann A, Kuster A, Dahmen H, Tacken I, Heinrich PC, Muller-Newen G (2000) Importance of the membrane-proximal extracellular domains for activation of the signal transducer glycoprotein 130. J Immunol 164:273–282

Lapidot T, Petit I (2002) Current understanding of stem cell mobilization: the roles of chemokines, proteolytic enzymes, adhesion molecules, cytokines, and stromal cells. Exp Hematol 30:973–981

Larsen A, Davis T, Curtis B, Gimpel S, Sims J, Cosman D, Park L, Sorensen E, March C, Smith C (1990) Expression cloning of a human granulocyte colony-stimulating factor receptor: a structural mosaic of hematopoietin receptor, immunoglobulin, and fibronectin domains. J Exp Med 172:1559–1570

Layton JE, Hall NE, Connell F, Venhorst J, Treutlein HR (2001) Identification of ligand-binding site III on the immunoglobulin-like domain of the granulocyte colony-stimulating factor receptor. J Biol Chem 276:36779–36787

Lee CK, Raz R, Gimeno R, Gertner R, Wistinghausen B, Takeshita K, DePinho RA, Levy DE (2002) STAT3 is a negative regulator of granulopoiesis but is not required for G-CSF-dependent differentiation. Immunity 17:63–72

Lieschke GJ (1997) CSF-deficient mice—what have they taught us?. [Review] [41 refs]. Ciba Found Symp 204:60–74

Lieschke GJ, Grail D, Hodgson G, Metcalf D, Stanley E, Cheers C, Fowler KJ, Basu S, Zhan YF, Dunn AR (1994) Mice lacking granulocyte colony-stimulating factor have chronic neutropenia, granulocyte and macrophage progenitor cell deficiency, and impaired neutrophil mobilization. Blood 84:1737–1746

Liu F, Wu HY, Wesselschmidt R, Kornaga T, Link DC (1996) Impaired production and increased apoptosis of neutrophils in granulocyte colony-stimulating factor receptor-deficient mice. Immunity 5:491–501

Liu Q, Sasaki T, Kozieradzki I, Wakeham A, Itie A, Dumont DJ, Penninger JM (1999) SHIP is a negative regulator of growth factor receptor-mediated PKB/Akt activation and myeloid cell survival. Genes Dev 13:786–791

Longley BJ, Reguera MJ, Ma Y (2001) Classes of c-KIT activating mutations: proposed mechanisms of action and implications for disease classification and therapy. Leuk Res 25:571–576

Lyman SD (1995) Biology of flt3 ligand and receptor. Int J Hematol 62:63–73

Marine JC, McKay C, Wang D, Topham DJ, Parganas E, Nakajima H, Pendeville H, Yasukawa H, Sasaki A, Yoshimura A, Ihle JN (1999) SOCS3 is essential in the regulation of fetal liver erythropoiesis. Cell 98:617–627

Masuhara M, Sakamoto H, Matsumoto A, Suzuki R, Yasukawa H, Mitsui K, Wakioka T, Tanimura S, Sasaki A, Misawa H, Yokouchi M, Ohtsubo M, Yoshimura A (1997) Cloning and characterization of novel CIS family genes. Biochem Biophys Res Commun 239:439–446

Matsumoto A, Masuhara M, Mitsui K, Yokouchi M, Ohtsubo M, Misawa H, Miyajima A, Yoshimura A (1997) CIS, a cytokine inducible SH2 protein, is a target of the JAK-STAT5 pathway and modulates STAT5 activation. Blood 89:3148–3154

Matsushita K, Arima N (1998) Involvement of granulocyte colony-stimulating factor in proliferation of adult T-cell leukemia cells. Leuk Lymphoma 31:295–304

McCracken S, Layton JE, Shorter SC, Starkey PM, Barlow DH, Mardon HJ (1996) Expression of granulocyte-colony stimulating factor and its receptor is regulated during the development of the human placenta. J Endocrinol 149:249–258

McCracken SA, Grant KE, MacKenzie IZ, Redman CW, Mardon HJ (1999) Gestational regulation of granulocyte-colony stimulating factor receptor expression in the human placenta. Biol Reprod 60:790–796

McLemore ML, Poursine-Laurent J, Link DC (1998) Increased granulocyte colony-stimulating factor responsiveness but normal resting granulopoiesis in mice carrying a targeted granulocyte colony-stimulating factor receptor mutation derived from a patient with severe congenital neutropenia. J Clin Invest 102:483–492

McLemore ML, Grewal S, Liu F, Archambault A, Poursine-Laurent J, Haug J, Link DC (2001) STAT-3 activation is required for normal G-CSF-dependent proliferation and granulocytic differentiation. Immunity 14:193–204

McPherson PS, Kay BK, Hussain NK (2001) Signaling on the endocytic pathway. Traffic 2:375–384

Medema RH, Kops GJ, Bos JL, Burgering BM (2000) AFX-like Forkhead transcription factors mediate cell-cycle regulation by Ras and PKB through p27kip1. Nature 404:782–787

Meraz MA, White JM, Sheehan KC, Bach EA, Rodig SJ, Dighe AS, Kaplan DH, Riley JK, Greenlund AC, Campbell D, Carver-Moore K, DuBois RN, Clark R, Aguet M, Schreiber RD (1996) Targeted disruption of the Stat1 gene in mice reveals unexpected physiologic specificity in the JAK-STAT signaling pathway. Cell 84:431–442

Metcalf D (1989) The molecular control of cell division, differentiation commitment and maturation in haemopoietic cells. Nature 339:27–30

Metcalf D (1991) Control of granulocytes and macrophages: molecular, cellular, and clinical aspects. Science 254:529–533

Miller WE, Lefkowitz RJ (2001) Expanding roles for β-arrestins as scaffolds and adapters in GPCR signaling and trafficking. Curr Opin Cell Biol 13:139–145

Molineux G, Pojda Z, Hampson IN, Lord BI, Dexter TM (1990) Transplantation potential of peripheral blood stem cells induced by granulocyte colony-stimulating factor. Blood 76:2153–2158

Moreno I, Martin G, Bolufer P, Barragan E, Rueda E, Roman J, Fernandez P, Le n P, Mena A, Cervera J, Torres A, Sanz MA (2003) Incidence and prognostic value of FLT3 internal tandem duplication and D835 mutations in acute myeloid leukemia. Haematologica 88:19–24

Morikawa K, Morikawa S, Miyawaki T, Nagasaki M, Torii I, Imai K (1996) Constitutive expression of granulocyte-colony stimulating factor receptor on a human B-lymphoblastoid cell line. Br J Haematol 94:250–257

Morikawa K, Morikawa S, Nakamura M, Miyawaki T (2002) Characterization of granulocyte colony-stimulating factor receptor expressed on human lymphocytes. Br J Haematol 118:296–304

Morishita K, Tsuchiya M, Asano S, Kaziro Y, Nagata S (1987) Chromosomal gene structure of human myeloperoxidase and regulation of its expression by granulocyte colony-stimulating factor. J Biol Chem 262:15208–15213

Murakami M, Narazaki M, Hibi M, Yawata H, Yasukawa K, Hamaguchi M, Taga T, Kishimoto T (1991) Critical cytoplasmic region of the interleukin 6 signal transducer gp130 is conserved in the cytokine receptor family. Proc Natl Acad Sci USA 88:11349–11353

Muszynski KW, Ruscetti FW, Heidecker G, Rapp U, Troppmair J, Gooya JM, Keller JR (1995) Raf-1 protein is required for growth factor-induced proliferation of hematopoietic cells. J Exp Med 181:2189–2199

Nagata S, Tsuchiya M, Asano S, Kaziro Y, Yamazaki T, Yamamoto O, Hirata Y, Kubota N, Oheda M, Nomura H, et al. (1986) Molecular cloning and expression of cDNA for human granulocyte colony-stimulating factor. Nature 319:415–418

Naka T, Narazaki M, Hirata M, Matsumoto T, Minamoto S, Aono A, Nishimoto N, Kajita T, Taga T, Yoshizaki K, Akira S, Kishimoto T (1997) Structure and function of a new STAT-induced STAT inhibitor. Nature 387:924–929

Nakajima H, Ihle JN (2001) Granulocyte colony-stimulating factor regulates myeloid differentiation through CCAAT/enhancer-binding protein epsilon. Blood 98:897–905

Nicholson SE, Oates AC, Harpur AG, Ziemiecki A, Wilks AF, Layton JE (1994) Tyrosine kinase JAK1 is associated with the granulocyte-colony-stimulating factor receptor and both become tyrosine-phosphorylated after receptor activation. Proc Natl Acad Sci USA 91:2985–2988

Nicholson SE, Willson TA, Farley A, Starr R, Zhang JG, Baca M, Alexander WS, Metcalf D, Hilton DJ, Nicola NA (1999) Mutational analyses of the SOCS proteins suggest a dual domain requirement but distinct mechanisms for inhibition of LIF and IL-6 signal transduction. EMBO J 18:375–385

Nicholson SE, De Souza D, Fabri LJ, Corbin J, Willson TA, Zhang JG, Silva A, Asimakis M, Farley A, Nash AD, Metcalf D, Hilton DJ, Nicola NA, Baca M (2000) Suppressor of cytokine signaling-3 preferentially binds to the SHP-2-binding site on the shared cytokine receptor subunit gp130. Proc Natl Acad Sci USA 97:6493–6498

Onishi M, Mui AL, Morikawa Y, Cho L, Kinoshita S, Nolan GP, Gorman DM, Miyajima A, Kitamura T (1996) Identification of an oncogenic form of the thrombopoietin receptor MPL using retrovirus-mediated gene transfer. Blood 88:1399–1406

Rausch O, Marshall CJ (1997) Tyrosine 763 of the murine granulocyte colony-stimulating factor receptor mediates Ras-dependent activation of the JNK/SAPK mitogen-activated protein kinase pathway. Mol Cell Biol 17:1170–1179

Rausch O, Marshall CJ (1999) Cooperation of p38 and extracellular signal-regulated kinase mitogen-activated protein kinase pathways during granulocyte colony-stimulating factor-induced hemopoietic cell proliferation. J Biol Chem 274:4096–4105

Roberts AW, Robb L, Rakar S, Hartley L, Cluse L, Nicola NA, Metcalf D, Hilton DJ, Alexander WS (2001) Placental defects and embryonic lethality in mice lacking suppressor of cytokine signaling 3. Proc Natl Acad Sci USA 98:9324–9329

Rodig SJ, Meraz MA, White JM, Lampe PA, Riley JK, Arthur CD, King KL, Sheehan KC, Yin L, Pennica D, Johnson EM, Jr., Schreiber RD (1998) Disruption of the Jak1 gene demonstrates obligatory and nonredundant roles of the Jaks in cytokine-induced biologic responses. Cell 93:373–383

Saito M, Yoshida K, Hibi M, Taga T, Kishimoto T (1992) Molecular cloning of a murine IL-6 receptor-associated signal transducer, gp130, and its regulated expression in vivo. J Immunol 148:4066–4071

Santini V, Scappini B, Indik ZK, Gozzini A, Rossi Ferrini P, Schreiber AD (2003) The Carboxyterminal region of the granulocyte-colony stimulating factor receptor transduces a phagocytic signal. Blood (in press)

Sato N, Asano S, Koeffler HP, Yoshida S, Takaku F, Takatani O (1988) Identification of neutrophil alkaline phosphatase-inducing factor in cystic fluid of a human squamous cell carcinoma as granulocyte colony-stimulating factor. J Cell Physiol 137:272–276

Schmitz J, Weissenbach M, Haan S, Heinrich PC, Schaper F (2000) SOCS3 exerts its inhibitory function on interleukin-6 signal transduction through the SHP2 recruitment site of gp130. J Biol Chem 275:12848–12856

Schnittger S, Schoch C, Dugas M, Kern W, Staib P, Wuchter C, Loffler H, Sauerland CM, Serve H, Buchner T, Haferlach T, Hiddemann W (2002) Analysis of FLT3 length mutations in 1003 patients with acute myeloid leukemia: correlation to cytogenetics, FAB subtype, and prognosis in the AMLCG study and usefulness as a marker for the detection of minimal residual disease. Blood 100:59–66

Schwaller J, Parganas E, Wang D, Cain D, Aster JC, Williams IR, Lee CK, Gerthner R, Kitamura T, Frantsve J, Anastasiadou E, Loh ML, Levy DE, Ihle JN, Gilliland DG (2000) Stat5 is essential for the myelo- and lymphoproliferative disease induced by TEL/JAK2. Mol Cell 6:693–704

Semerad CL, Liu F, Gregory AD, Stumpf K, Link DC (2002) G-CSF is an essential regulator of neutrophil trafficking from the bone marrow to the blood. Immunity 17:413

Sherr CJ, Roberts JM (1995) Inhibitors of mammalian G1 cyclin-dependent kinases. Genes Dev 9:1149–1163

Shimoda K, Okamura S, Harada N, Kondo S, Okamura T, Niho Y (1993) Identification of a functional receptor for granulocyte colony-stimulating factor on platelets. J Clin Invest 91:1310–1313

Shimoda K, Iwasaki H, Okamura S, Ohno Y, Kubota A, Arima F, Otsuka T, Niho Y (1994) G-CSF induces tyrosine phosphorylation of the JAK2 protein in the human myeloid G-CSF responsive and proliferative cells, but not in mature neutrophils. Biochem Biophys Res Commun 203:922–928

Shimoda K, Feng J, Murakami H, Nagata S, Watling D, Rogers NC, Stark GR, Kerr IM, Ihle JN (1997) Jak1 plays an essential role for receptor phosphorylation and Stat activation in response to granulocyte colony-stimulating factor. Blood 90:597–604

Shimozaki K, Nakajima K, Hirano T, Nagata S (1997) Involvement of STAT3 in the granulocyte colony-stimulating factor-induced differentiation of myeloid cells. J Biol Chem 272:25184–25189

Shultz LD, Schweitzer PA, Rajan TV, Yi T, Ihle JN, Matthews RJ, Thomas ML, Beier DR (1993) Mutations at the murine motheaten locus are within the hematopoietic cell protein-tyrosine phosphatase (Hcph) gene. Cell 73:1445–1454

Sinha S, Jancarik J, Roginskaya V, Rothermund K, Boxer LM, Corey SJ (2001) Suppression of apoptosis and granulocyte colony-stimulating factor-induced differentiation by an oncogenic form of Cbl. Exp Hematol 29:746–755

Souza LM, Boone TC, Gabrilove J, Lai PH, Zsebo KM, Murdock DC, Chazin VR, Bruszewski J, Lu H, Chen KK, et al. (1986) Recombinant human granulocyte colony-stimulating factor: effects on normal and leukemic myeloid cells. Science 232:61–65

Starr R, Willson TA, Viney EM, Murray LJ, Rayner JR, Jenkins BJ, Gonda TJ, Alexander WS, Metcalf D, Nicola NA, Hilton DJ (1997) A family of cytokine-inducible inhibitors of signaling. Nature 387:917–921

Takahashi Y, Carpino N, Cross JC, Torres M, Parganas E, Ihle JN (2003) SOCS3: an essential regulator of LIF receptor signaling in trophoblast giant cell differentiation. EMBO J 22:372–384

Tapley P, Shevde NK, Schweitzer PA, Gallina M, Christianson SW, Lin IL, Stein RB, Shultz LD, Rosen J, Lamb P (1997) Increased G-CSF responsiveness of bone marrow cells from hematopoietic cell phosphatase deficient viable motheaten mice. Exp Hematol 25:122–131

Teglund S, McKay C, Schuetz E, van Deursen JM, Stravopodis D, Wang D, Brown M, Bodner S, Grosveld G, Ihle JN (1998) Stat5a and Stat5b proteins have essential and nonessential, or redundant, roles in cytokine responses. Cell 93:841–850

Tenen DG (2001) Abnormalities of the CEBP α transcription factor: a major target in acute myeloid leukemia. Leukemia 15:688–689

Tenen DG, Hromas R, Licht JD, Zhang DE (1997) Transcription factors, normal myeloid development, and leukemia. Blood 90:489–519

Thomas J, Liu F, Link DC (2002) Mechanisms of mobilization of hematopoietic progenitors with granulocyte colony-stimulating factor. Curr Opin Hematol 9:183–189

Tian SS, Lamb P, Seidel HM, Stein RB, Rosen J (1994) Rapid activation of the STAT3 transcription factor by granulocyte colony-stimulating factor. Blood 84:1760–1764

Tian SS, Tapley P, Sincich C, Stein RB, Rosen J, Lamb P (1996) Multiple signaling pathways induced by granulocyte colony-stimulating factor involving activation of JAKs, STAT5, and/or STAT3 are required for regulation of three distinct classes of immediate early genes. Blood 88:4435–4444

Tsuchiya H, el-Sonbaty SS, Watanabe M, Suzushima H, Asou N, Murakami T, Takeda T, Shimosaka A, Takatsuki K, Matsuda I (1993) Analysis of myeloid characteristics in acute lymphoblastic leukemia. Leuk Res 17:809–813

Tsui HW, Siminovitch KA, de Souza L, Tsui FW (1993) Motheaten and viable motheaten mice have mutations in the haematopoietic cell phosphatase gene. Nat Genet 4:124–129

van der Geer P, Wiley S, Gish GD, Pawson T (1996) The Shc adaptor protein is highly phosphorylated at conserved, twin tyrosine residues (Y239/240) that mediate protein–protein interactions. Curr Biol 6:1435–1444

Vogel W, Ullrich A (1996) Multiple in vivo phosphorylated tyrosine phosphatase SHP-2 engages binding to Grb2 via tyrosine 584. Cell Growth Differ 7:1589–1597

Ward AC, Hermans MH, Smith L, van Aesch YM, Schelen AM, Antonissen C, Touw IP (1999a) Tyrosine-dependent and -independent mechanisms of STAT3 activation by the human granulocyte colony-stimulating factor (G-CSF) receptor are differentially utilized depending on G-CSF concentration. Blood 93:113–124

Ward AC, Smith L, de Koning JP, van Aesch Y, Touw IP (1999b) Multiple signals mediate proliferation, differentiation, and survival from the granulocyte colony-stimulating factor receptor in myeloid 32D cells. J Biol Chem 274:14956–14962

Ward AC, van Aesch YM, Gits J, Schelen AM, de Koning JP, van Leeuwen D, Freedman MH, Touw IP (1999c) Novel point mutation in the extracellular domain of the granulocyte colony-stimulating factor (G-CSF) receptor in a case of severe congenital neutropenia hyporesponsive to G-CSF treatment. J Exp Med 190:497–508

Ward AC, van Aesch YM, Schelen AM, Touw IP (1999d) Defective internalization and sustained activation of truncated granulocyte colony-stimulating factor receptor found in severe congenital neutropenia/acute myeloid leukemia. Blood 93:447–458

Ward AC, Loeb DM, Soede-Bobok AA, Touw IP, Friedman AD (2000a) Regulation of granulopoiesis by transcription factors and cytokine signals. Leukemia 14:973–990

Ward AC, Oomen SP, Smith L, Gits J, van Leeuwen D, Soede-Bobok AA, Erpelinck-Verschueren CA, Yi T, Touw IP (2000b) The SH2 domain-containing protein tyrosine phosphatase SHP-1 is induced by granulocyte colony-stimulating factor (G-CSF) and modulates signaling from the G-CSF receptor. Leukemia 14:1284–1291

Wells JA, de Vos AM (1996) Hematopoietic receptor complexes. Annu Rev Biochem 65:609–634

Welte K, Boxer LA (1997) Severe chronic neutropenia: pathophysiology and therapy. Semin Hematol 34:267–278

Welte K, Zeidler C, Reiter A, Muller W, Odenwald E, Souza L, Riehm H (1990) Differential effects of granulocyte-macrophage colony-stimulating factor and granulocyte colony-stimulating factor in children with severe congenital neutropenia. Blood 75:1056–1063

White SM, Ball ED, Ehmann WC, Rao AS, Tweardy DJ (1998) Increased expression of the differentiation-defective granulocyte colony-stimulating factor receptor mRNA isoform in acute myelogenous leukemia. Leukemia 12:899–906

White SM, Alarcon MH, Tweardy DJ (2000) Inhibition of granulocyte colony-stimulating factor-mediated myeloid maturation by low level expression of the differentiation-defective class IV granulocyte colony-stimulating factor receptor isoform. Blood 95:3335–3340

Yasukawa H, Misawa H, Sakamoto H, Masuhara M, Sasaki A, Wakioka T, Ohtsuka S, Imaizumi T, Matsuda T, Ihle JN, Yoshimura A (1999) The JAK-binding protein JAB inhibits Janus tyrosine kinase activity through binding in the activation loop. EMBO J 18:1309–1320

Yasukawa H, Sasaki A, Yoshimura A (2000) Negative regulation of cytokine signaling pathways. Annu Rev Immunol 18:143–164

Yi TL, Cleveland JL, Ihle JN (1992) Protein tyrosine phosphatase containing SH2 domains: characterization, preferential expression in hematopoietic cells, and localization to human chromosome 12p12-p13. Mol Cell Biol 12:836–846

Zhan Y, Lieschke GJ, Grail D, Dunn AR, Cheers C (1998) Essential roles for granulocyte-macrophage colony-stimulating factor (GM-CSF) and G-CSF in the sustained hematopoietic response of Listeria monocytogenes-infected mice. Blood 91:863–869

Zhang JG, Farley A, Nicholson SE, Willson TA, Zugaro LM, Simpson RJ, Moritz RL, Cary D, Richardson R, Hausmann G, Kile BJ, Kent SB, Alexander WS, Metcalf D, Hilton DJ, Nicola NA, Baca M (1999) The conserved SOCS box motif in suppressors of cytokine signaling binds to elongins B and C and may couple bound proteins to proteasomal degradation. Proc Natl Acad Sci USA 96:2071–2076

T. Hanada · I. Kinjyo · K. Inagaki-Ohara · A. Yoshimura

Negative regulation of cytokine signaling by CIS/SOCS family proteins and their roles in inflammatory diseases

Published online: 29 March 2003
© Springer-Verlag 2003

Abstract Immune and inflammatory systems are controlled by multiple cytokines, including interleukins (ILs) and interferons. These cytokines exert their biological functions through *Janus* tyrosine kinases (JAKs) and STAT transcription factors. The CIS (cytokine-inducible SH2 protein) and SOCS (suppressors of cytokine signaling) are a family of intracellular proteins, several of which have emerged as key physiological regulators of cytokine responses, including those that regulate the inflammatory systems. In this review, we focused on the molecular mechanism of the action of CIS/SOCS family proteins and their roles in inflammatory diseases. Furthermore, we illustrate several approaches for treating inflammatory diseases by modulating extracellular and intracellular signaling pathways.

Introduction

The inflammatory response consists of the sequential release of mediators, including inflammatory cytokines and the recruitment of circulating leukocytes, which become activated at the inflammatory site and release further mediators. However, in most cases, the inflammatory response is resolved by the release of endogenous anti-inflammatory mediators (anti-inflammatory cytokines), as well as the accumulation of intracellular negative regulatory factors. Thus, the inflammatory cells are cleared at an appropriate time. However, the persistent accumulation and activation of leukocytes are a hallmark of chronic inflammation, suggesting a dysfunction of these negative regulatory mechanisms. Current clinical approaches to the treatment of inflammation mostly focus on the inhibition of proinflammatory mediator production and the suppression of the initiation of the inflammatory response, i.e., the suppression of positive signaling pathways of proinflammatory

T. Hanada · I. Kinjyo · K. Inagaki-Ohara · A. Yoshimura (✉)
Division of Molecular and Cellular Immunology, Medical Institute of Bioregulation, Kyushu University, Maidashi, 812-8582 Higashi-ku, Fukuoka, Japan
e-mail: yakihiko@bioreg.kyushu-u.ac.jp · Tel.: +81-92-6426823 · Fax: +81-92-6426825

cytokines. However, the mechanisms by which the inflammatory response is resolved might provide new targets in the treatment of inflammatory diseases.

Well-characterized inflammatory cytokines are interleukin-1 (IL-1), tumor necrosis factor-α (TNF-α), γ-interferon (IFN-γ), IL-12, IL-18, and granulocyte-macrophage colony-stimulating factor (GM-CSF), while anti-inflammatory cytokines are IL4, IL-10, IL-13, IFN-γ, and transforming growth factor (TGF)-β. The intracellular signal transduction pathways of these cytokines have been studied extensively, and these pathways ultimately activate transcription factors, such as NF-kB (IL-1, IL-18 and TNF-α), Smad (TGF-β), and STATs (IL-6, IL-12, IL-10, and IFN-γ). Recently, the negative-feedback regulation of these pathways has been identified and shown to be very important for immune and inflammatory regulation. Among them, the CIS/SOCS family, which mainly regulates the JAK/STAT pathway, is the paradigm of such negative feedback regulation.

JAK/STAT pathway

Cytokines including interleukins, interferons, and hematopoietins are structurally related and modulate immunity and inflammation. The receptors for this class of cytokines form a receptor family which is characterized by conserved extracellular domains that include the Trp-Ser-Xaa-Trp-Ser pentapeptide motif (where Xaa is any amino acid; Gearing et al. 1989). Signaling from cytokine receptors is initiated by receptor oligomerization that is induced by cytokine binding, which brings associated JAK kinases (JAK1, JAK2, JAK3, and Tyk2) into close apposition and allows their cross-phosphorylation and activation (Fig. 1; Ihle 1995). The activated JAKs phosphorylate the receptor cytoplasmic domains, which creates docking sites for SH2-containing signaling proteins. Among the substrates of tyrosine phosphorylation are members of the signal transducers and activators of the transcription family of proteins (STATs; Ihle 1996; Darnell 1997). Although this pathway was initially found to be activated by IFNs, it is now known that a large number of cytokines, growth factors, and hormonal factors activate JAK and/or STAT proteins. For example, proinflammatory cytokine IL-6 binds to the IL-6 receptor α chain and gp130, which mainly activate JAK1 and STAT3. IFN-γ utilizes JAK1 and JAK2 and mainly activates STAT1. Interestingly, anti-inflammatory cytokine IL-10 also activates STAT3 (O'Farrell et al. 1998). STAT4 and STAT6 are essential for Th1 and Th2 development since these are activated by IL-12 and IL-4, respectively (Kaplan et al. 1996; Shimoda et al. 1996; Takeda et al. 1996).

The JAKs and STATs are essential intracellular mediators of immune cytokine action, which is probed by gene-knockout mice (Ivashkiv 2000). Nevertheless, control of the magnitude and duration of signaling is also essential to prevent pathology (Duhe et al. 2001; Yasukawa et al. 2000). Receptor internalization, tyrosine phosphatases, and members of the protein inhibitors of activated STAT (PIAS) family all contribute to this negative regulatory network (Liu et al. 2001). However, those mechanism are relatively non-specific; tyrosine-phosphatases can downregulate any growth-factor-induced signals, and PIASs has been shown to act as a general transcriptional regulator by sumoylation (Jackson 2001). The discovery of the SOCS proteins has defined an important central mechanism for the negative regulation of the JAK–STAT pathway (Fig. 1).

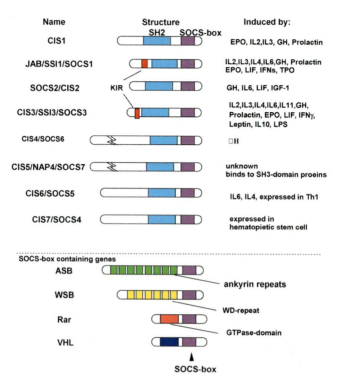

Fig. 1 The alternative names and domain structure of the SOCS protein family. The kinase inhibitory region (*KIR*) of SOCS1 and SOCS3 is indicated in *red*. SOCS-box containing proteins are also listed. The alternative nomenclature for each SOCS protein is given in *parentheses*. *CIS* cytokine-induced SH2 protein, *JAB* Janus kinase (JAK)-binding protein, *NAP4* Nck, Ash and phospholipase-C binding protein, *SH2* SRC-homology 2, *SOCS* suppressor of cytokine signaling, *SSI* STAT-induced STAT inhibitor, *WSB* WD-40 repeats-SOCS-box, *ASB* ankyrin repeats-SOCS-box

Negative regulation of the JAK/STAT pathway by the CIS/SOCS family

The longevity of cytokine signals transduced by the JAK/STAT pathway is regulated, in part, by a family of endogenous JAK kinase inhibitor proteins referred to as suppressors of cytokine signaling (SOCS) or cytokine-inducible SH2 proteins (CIS; Krebs and Hilton 2001; Hanada and Yoshimura 2002). The first identified CIS/SOCS gene, CIS1, has been shown to be a negative-feedback regulator of the STAT5 pathway (Yoshimura et al. 1995; Matsumoto et al. 1999). CIS1 binds to the phosphorylated tyrosine residues of cytokine receptors such as the EPO receptor, IL-3 receptor β chain, IL-2 receptor β chain (Aman et al. 1999), growth hormone receptor (Hansen et al. 1999), and prolactin receptor (Pezet et al. 1999; Tonko-Geymayer 2002; Endo et al. 2003) through the SH2 domain, thereby masking STAT5 docking sites. All these receptors commonly activate STAT5 and CIS1 does not bind to cytokine receptors that activate STAT1 and STAT3, including gp130. Therefore, CIS1 is a very specific negative regulator of STAT5 (Yoshimura et al. 1995). This was confirmed in vivo by generating transgenic mouse of CIS1. This transgenic

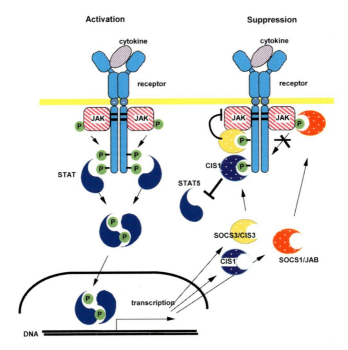

Fig. 2 The molecular mechanism by which SOCS proteins negatively regulate cytokine signaling. Cytokine stimulation activates the JAK-STAT pathway, leading to the induction of CIS, SOCS1, and/or SOCS3. CIS, SOCS1, and SOCS3 appear to inhibit signaling by different mechanisms: SOCS1 binds to the JAKs and inhibits catalytic activity, SOCS3 binds to JAK-proximal sites on cytokine receptors and inhibits JAK activity, and CIS blocks the binding of STATs to cytokine receptors

mouse ubiquitously expressing CIS1 showed similar phenotypes of STAT5-knockout, such as growth retardation as well as suppression of mammary grand development and T-cell proliferation (Matsumoto et al. 1999). The phenotype of CIS1 gene knockout mice has not been published, but at least in C57BL/6 background, we observed significant defects in T cells (unpublished observations). Therefore, CIS1 seems to be very important for immunoregulation.

We and others recently cloned other CIS family members, SOCS1/JAB (JAK-binding protein) which directly bind to the JAK2 tyrosine kinase domain and inhibit JAK tyrosine kinase activity (Endo et al. 1997; Naka et al. 1997; Starr et al. 1997). At present, the SOCS family contains eight members of related proteins that share a common modular organization of an SH2 domain followed by a short motif called SOCS-box (Masuhara et al. 1997; Hilton et al. 1998). Although there is sequence homology between all family members (particularly in the SOCS box and SH2 domain), CIS and SOCS2, SOCS1 and SOCS3, SOCS4 and SOCS5, and SOCS6 and SOCS7 have marked pair-wise homology across the entire protein sequence (Fig. 1).

Both SOCS1 and SOCS3 inhibit JAK tyrosine kinase activity; SOCS1 directly binds to the activation loop of JAKs through the SH2 domain, while SOCS3 binds to the cytokine receptors (Fig. 2). These two molecules contain a similar kinase inhibitor region (KIR) at the N-terminus that is essential for JAK inhibition (Yasukawa et al. 1999; Sasaki et al. 1999). We proposed that KIR interacts with the region close to the catalytic groove of the

JAK2 kinase domain, thereby preventing the access of substrates to the catalytic pocket. SOCS3 has been shown to bind to Y757 of gp130, Y985 of the leptin receptor, as well as to Y401 of the EPO receptor, some of which are the same binding sites for protein tyrosine phosphatase 2 (SHP-2; Schmitz et al. 2000; Nicholson et al. 2000; Sasaki et al. 2000; Hortner et al. 2002). As SHP2 can promote gp130 signaling through the activation of mitogen-activated protein kinases, it is possible that SOCS3 might suppress aspects of gp130 signaling also by competing with SHP2 for receptor binding.

Function of the SOCS-box

The conserved SOCS-box domain interacts with Elongins B and C/Cul2/Rbx-1 proteins that form part of an E3 ubiquitin ligase complex (Kamura et al. 1998; Zhang et al. 1999) that ubiquitinates and targets associated proteins (or SOCS proteins themselves) for degradation. The Elongin BC complex was identified initially to be a positive regulator of RNA polymerase II elongation factor Elongin A (Bradsher et al. 1993; Aso et al. 1995) and subsequently as a component of the multiprotein von Hippel-Lindau disease (VHL) tumor-suppressor complex (Duan et al. 1995). The VHL protein also contains a SOCS-box-like motif and mediates ubiquitination and degradation of the HIF transcription factor that interacts with the N-terminal region of VHL (Kibel et al. 1995). Therefore, like VHL, the SOCS proteins seem to have a generic mechanism that targets these components for ubiquity-mediated proteasome degradation. Indeed, the activation of JAKs and STATs can be prolonged in the presence of proteasome inhibitors (Yu and Burakoff 1997; Verdier et al. 1988), and the interactions of CIS with the erythropoietin receptor (EPOR) and of SOCS1 with VAV and IRS-1, -2 seem to promote the proteasomal degradation of these proteins (Verdier et al. 1988; De Sepulveda et al. 2000; Rui et al. 2002). Furthermore, we and others reported that the expression of SOCS1 in cells that have been transformed by the TEL–JAK2 fusion protein, an active form of JAK2 found in T-cell leukemias, suppresses factor-dependent growth and tumorigenicity. Although SOCS1 inhibited the catalytic activity of the fusion protein, full suppression relied on the SOCS-box-dependent ubiquitination of TEL–JAK2 (Kamizono et al. 2001; Frantsve et al. 2001). Endogenous JAK2 was also shown to be ubiquitinated and degraded by SOCS1 in a SOCS-box-dependent manner (Ungureanu et al. 2002). Although definitive proof that SOCS proteins direct ubiquitin-mediated proteasomal degradation of associated proteins in a SOCS-box-dependent manner in vivo is still lacking, recent data from mice that were genetically modified to lack only the SOCS box of SOCS1 confirm that this domain is crucial for the complete in vivo suppression of cytokine signaling (Zhang et al. 2001).

Related to the function of the SOCS-box, we have found that the transgenic (Tg) mice expressing a mutant SOCS1 in the KIR region (F59D-JAB) exhibited a more potent STAT3 activation and a more severe colitis than did wild-type littermates after treatment with dextran sulfate sodium. We now find that there is a prolonged activation of JAKs and STATs in response to a number of cytokines in T cells from Tg mice with lck promoter-driven F59D-JAB (Hanada et al. 2001). We found that C-terminal SOCS-box played an essential role in augmenting cytokine signaling by F59D-JAB. It has been reported that interaction between the SOCS-box and the Elongin BC complex stabilizes SOCS1. F59D-JAB induced destabilization of wild-type SOCS1, while overexpression of Elongin BC canceled this effect. Levels of endogenous SOCS1 and SOCS3 in T cells from F59D-JAB Tg-mouse were lower than in wild type mice. Therefore, we propose that F59D-JAB destabilizes wild-type, endogenous SOCSs by chelating the Elongin BC complex, thereby sustaining JAK activation. In this case, SOCS-box is necessary for the protein stability of SOCS molecules. Further study is necessary to define the function of the SOCS box.

STAT and inflammation

Aberrant expression of LIF/IL-6 family cytokines has been associated with autoimmune disease, septic shock, and neoplasia (Hanada and Yoshimura 2002; Hirano et al. 1990). Constitutive activation of STAT3 was often observed in chronic inflammation, probably reflecting high levels of IL-6. We have shown constitutive phosphorylation STAT3 among six STAT species in IBD and RA patients as well as experimental IBD and RA models in mice. Elevated STAT1 activation is also observed in asthma patients' epithelial cells (Sampath et al. 1999). STAT4 transgenic mice also develop colitis (Wirtz et al. 1999), and IL-12/STAT-4-driven Th1 responses have been shown to predominate in human Crohn's disease (Parrello et al. 2000). Therefore, STATs activation usually plays a positive role in inflammation.

To assess the role of STAT3 in vivo more precisely, the STAT3 gene was disrupted in a tissue- or cell-specific manner by the Cre-loxP recombination system. In STAT3-deficient T cells, IL-6-induced T-cell proliferation was impaired due to the lack of IL-6-mediated prevention of apoptosis (Takeda et al. 1998), which is consistent with the protective effect of the anti-IL-6 receptor monoclonal antibody against T cell-mediated colitis and inflammatory arthritis models (Yamamoto et al. 2000; Atreya et al. 2000). Usually, activation of STAT3 induces proliferation and antiapoptosis through induction of pim-1, c-myc, cyclin-D, and Bcl-X (Shirogane et al. 1999). STAT3 also promotes hyperplasia in late phase of inflammation because it promotes cell proliferation. We have shown that IL-6 produced in synoviocytes from RA patients functions as an autocrine growth factor. These data suggest that activation of STAT3 participates in the development of inflammation through hyperplasia of epithelial cells and fibroblasts and the survival of activated T cells. STAT3 activation may also have a prominent role in promoting inflammation by enhancing inflammatory cytokine production from these cells.

However, STAT3 in macrophages apparently plays a protective role from inflammation. Takeda et al. showed that conditional knockout of STAT3 in macrophages and neutrophils resulted in chronic enterocolitis with age (Takeda et al. 1999). This is probably due to the enhancement of the Th1 response by the block of anti-inflammatory cytokine IL-10 signaling, which utilizes STAT3. Thus, STAT3 may play a role of protection from tissue damage in acute inflammation or recovery of tissue injury (Tebbutt et al. 2002; Hong et al. 2002). Thus, STAT3 seems to play positive and negative roles in inflammation, depending on tissues, cause, and phase. On the other hand, STAT1 promotes inflammation by inducing apoptosis of tissues (Hong et al. 2002).

SOCS1 and inflammation

The study of SOCS1 knockout mice revealed that SOCS1 is essential for IFN-γ signal suppression and T-cell activation. Although SOCS1 KO mice are normal at birth, they exhibit stunted growth and die within 3 weeks of age with a syndrome characterized by severe lymphopenia, activation of peripheral T cells, fatty degeneration and necrosis of the liver, and macrophage infiltration of major organs (Naka et al. 1998; Starr et al. 1998). The neonatal defects exhibited by SOCS1$^{-/-}$ mice appear to occur primarily as a result of unbridled IFN-γ signaling, since SOCS1$^{-/-}$ mice that also lack the IFN-γ gene do not die neonatally (Alexander et al. 1999; Marine et al. 1999). Constitutive activation of STAT1 as well as constitutive expression of IFN-γ-inducible genes was observed in SOCS1 KO mice. In mice, SOCS1 is expressed predominately in T cells, but IFN-γ induced SOCS1 in most types of cells. SOCS1$^{-/-}$ mice that also lack the Rag2 gene and therefore lack functional lymphocytes also survived (Marine et al. 1999). Furthermore, reconstitution of the lymphoid lineage of irradiated JAK3$^{-/-}$ mice with SOCS1$^{-/-}$ bone marrow recapitulated the same fatal syndrome (Marine et al. 1999). These data strongly suggest that the excess IFN-γ is derived from the abnormally activated T cells in SOCS1$^{-/-}$ mice. Although neonatal or early adult disease was avoided by removing IFN-γ, loss of SOCS1 significantly shortened the lifespan of the mice. The major causes of premature death were the development of polycystic kidneys, pneumonia, chronic skin ulcers, and chronic granulomas in the gut and various other organs (Metcalf et al. 2002), while SOCS1$^{-/-}$/IFN-$\gamma^{+/-}$ mice develop an autoimmune polymyositis about 160 days after birth (Metcalf et al. 2000).

Recently, Ernst et al. generated unique mice with a COOH-terminal truncated gp130-STAT "knock-in" mutation which deleted all STAT-binding sites (Ernst et al. 2001). Unlike mice with null mutations in any of the components in the gp130 signaling pathway, gp130-STAT mice displayed gastrointestinal ulceration and a severe joint disease with features of chronic synovitis, cartilaginous metaplasia, and degradation of the articular cartilage. They found that mitogenic hyperresponsiveness of synovial cells to the LIF/IL-6 family of cytokines was caused by sustained gp130-mediated SHP-2/ras/Erk activation (Fig. 1) due to impaired STAT-mediated induction of SOCS1, which normally limits gp130 signaling. These data strongly suggest that imparted SOCS1 induction in tissues is susceptible to inflammation. However, little is known about the expression of SOCS-1 in human chronic inflammatory diseases.

SOCS1 is also implicated in liver inflammation and hepatocarcinoma development. SOCS1-deficient NK- and NKT cells were spontaneously activated and induce damage of the liver (Naka et al. 2001). We found that SOCS$^{-/+}$ mice showed stronger liver damage in response to chemicals (T. Yoshida et al., unpublished data). Thus, reduced expression of SOCS1 could be a risk for hepatitis and hepatoma. Expression of the SOCS1 gene has been shown to be repressed by DNA methylation in hepatocarcinoma (Nagai et al. 2002; Yoshikawa et al. 2001).

SOCS1 and SOCS3 are shown to be induced in concanavalin A (Con A)-induced, T cell-mediated hepatitis (Hong et al. 2002). In this model, STAT1 induces expression of SOCS1, resulting in the suppression of STAT3-controlled antiapoptotic signals. On the other hand, IL-6 induced STAT3 activation, resulting in upregulation of SOCS3, which suppresses STAT1-induced proapoptotic signals and therefore ameliorates liver injury. Thus, STAT1 and STAT3 in hepatocytes negatively regulate one another through the induction of SOCS.

SOCS3 and inflammation

SOCS3 knockout mice die during the embryonic stage of development either by dysregulated fetal liver erythropoiesis or defects of placenta functions (Marine et al. 1999; Roberts

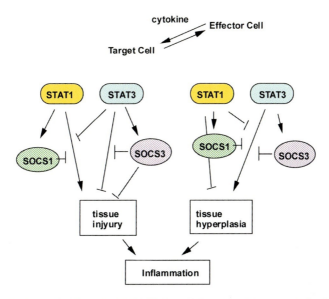

Fig. 3 Current simple model of the roles of STAT1,3 and SOCS1,3 in inflammation. STAT1 promotes cell apoptosis and growth inhibition, while STAT3 functions antiapoptosis and stimulates cell-proliferation. Therefore STAT1 promotes tissue injury but suppresses tissue hyperplasia, while STAT3 is necessary for the prevention of tissue damage but promotes hyperplasia. SOCS1 is mainly induced through STAT1 and SOCS3 is induced by STAT3. They function for negative feedback as well as for cross-suppression of different STATs. Therefore, SOCSs can regulate inflammation differently, depending on cytokines and type of inflammation. More complicatedly, STATs and SOCSs have different roles in effector cells (lymphocytes and macrophages)

et al. 2001). However, the physiological function of SOCS3 in adult tissues remains to be determined. Many reports have indicated that SOCS3 is induced by various inflammatory and anti-inflammatory cytokines such as IL-6, IL-12, IFN-γ, and IL-10 and that it negatively regulates those cytokine actions as well as STAT functions (Cassatella et al. 1999). Moreover, SOCS3 has been shown to be induced by IL-1 and TNF-α, as well as LPS (Boisclair et al. 2000; Bode et al. 1999). Thus, we examined the expression of SOCS3 in human chronic inflammatory diseases such as IBD and RA. We found that SOCS3 was highly expressed in epithelial and lamina propria cells in the colon of IBD model mice as well as human UC and CD patients (Suzuki et al. 2001). In a DSS-induced mouse colitis model, a time-course experiment indicated that STAT3 activation was 1 day ahead of SOCS3 induction; STAT3 activation became apparent during days 3–5 and decreased thereafter, while SOCS3 expression was induced at day 5 and maintained high levels thereafter. High levels of SOCS3 expression were also observed in human RA but not in OA patients (Shouda et al. 2001). In murine models of inflammatory synovitis, STAT3 phosphorylation preceded SOCS-3 expression, which is consistent with the idea that SOCS-3 is part of a JAK/STAT negative-feedback loop (Zhang et al. 1999; Suzuki et al. 2001). Based on the evidence that forced expression of SOCS3 can inhibit IL-6-mediated STAT3 activation, SOCS3, which is induced by STAT3 activation, acts as a negative-feedback regulator of STAT3. These data raise the possibility that SOCS3 expression is one, if not the only, mechanism that negatively regulates inflammatory reaction in colitis and arthritis. Our current model of the role of STAT and SOCS in inflammation is shown in Fig. 3.

A mouse line of mutated gp130, in which the SHP-2/SOCS3 binding site was disrupted, developed a rheumatoid arthritis (RA)-like joint disease with increased production of Th1-type cytokines and Igs of the IgG2a and IgG2b classes (Atsumi et al. 2002). Therefore, SOCS3 is also a key factor in the development of autoimmune disease.

SOCS and T-cell regulation

Apparently, SOCS1 is an important anti-inflammatory gene, judged from KO mice phenotype. Inflammation in SOCS1 KO mice is apparently dependent on T cells. SOCS1$^{-/-}$ T cells as well as NKT are activated, and therefore injure their own tissues. Thus, SOCS1$^{-/-}$ mice have some similarity to autoimmune diseases. However, the activation mechanism of SOCS1$^{-/-}$ T cells has not been clarified. Naka et al. demonstrated that the major cause of death of SOCS1$^{-/-}$ mice is liver injury induced by NKT cells (Naka et al. 2001). In this case, SOCS1 also functions as a negative regulator of IL-4/STAT6 signaling. Recently, another group showed that SOCS1 negatively regulates IL-12 (Eyles et al. 2002). Therefore, SOCS1 plays an essential role in regulating cytokine signals in T cells, and is probably necessary for maintaining the anergic phenotype of T cells (McHugh et al. 2002).

Unlike wild-type splenocytes, which require both interleukin-2 (IL-2) and T-cell receptor (TCR) ligation for significant proliferation in vitro, cells from SOCS1$^{-/-}$ spleens proliferate strongly in response to IL-2 alone (Marine et al. 1999). Augmented proliferative responses of SOCS1$^{-/-}$ thymocytes to IL-4 have been observed (Naka et al. 2001), and mice that lack both SOCS1 and STAT6, which mediates IL-4 signaling, have delayed onset of mortality compared with mice that lack SOCS1 only, which indicates that IL-4 might contribute to disease (Marine et al. 1999). A recent report has shown that IL-6 fails to block IFN-γ production and signaling in SOCS1$^{-/-}$ CD4$^+$ T cells, which indicates a mechanism by which SOCS1 might modulate the inhibition of T helper 1 (TH1) differentiation by IL-6 (Diehl et al. 2000), and increased apoptotic responses of SOCS1$^{-/-}$ cells to TNF-α have also been reported (Morita et al. 2000). SOCS1 can also interact with components of the TCR and seems to be able to block TCR signaling (Matsuda et al. 2000). It is anticipated that roles for these and other signal-transduction systems in SOCS1$^{-/-}$ disease, as well as other actions of SOCS1 in T-cell development, regulation, and immune responses will continue to be actively pursued.

SOCS3 also modulates T-cell response. SOCS-3 is shown to inhibit transcription driven by the IL-2 promoter in response to T-cell activation. This inhibitory activity correlates with the suppression of calcineurin-dependent dephosphorylation and activation of the IL-2 promoter binding transcription factor, NFATp, by binding and inhibiting the catalytic subunit of calcineurin (Banerjee et al. 2002). SOCS3 is also implicated in Th2 development. SOCS3 mRNA was selectively expressed in Th2 cells but not in Th1 cells (Egwuagu et al. 2002). Therefore, SOCS3 may be involved in Th2-type diseases like allergy and asthma. On the contrary, SOCS5 was predominantly expressed in Th1 cells (Seki et al. 2002). We found that SOCS5 interacted with the cytoplasmic region of the IL-4 receptor α chain irrespective of receptor tyrosine phosphorylation. This unconventional interaction of SOCS5 protein with the IL-4 receptor resulted in the inhibition of IL-4-mediated STAT6-6 activation. Therefore, the induced SOCS5 protein in a Th1 differentiation environment may play an important role by regulating Th1 and Th2 balance.

SOCS and TLR signaling

Bacterial lipopolysaccharide (LPS) triggers innate immune responses through Toll-like receptor (TLR) 4. Other bacterial pathogens, including CpG-DNA, activate TLR family receptors (see review Akira et al. 2001). Regulation of TLR signaling is a key step for inflammation, septic shock, and innate/adaptive immunity. SOCS1 and SOCS3 were found to be induced by LPS or CpG-DNA stimulation in macrophages (Stoiber et al. 1999; Crepso et al. 2000; Dalpke et al. 2001). SOCS1 has been implicated in the hyporesponsiveness to cytokines such as IFN-γ after exposure of LPS to macrophages (Crespo et al. 2002). On the other hand, we and others found that SOCS1-deficient mice were more sensitive to LPS shock than wild-type littermates (Kinjyo et al. 2002; Nakagawa et al. 2002). SOCS1 is expressed in macrophages following LPS-stimulation. Nakagawa et al. showed that SOCS1$^{-/-}$ mice (predisease onset), SOCS1$^{+/-}$ as well as IFN-$\gamma^{-/-}$SOCS1$^{-/-}$ mice were hyperresponsive to LPS, and were very sensitive to LPS-induced lethality. Macrophages from these mice produced increased levels of the proinflammatory cytokines TNF-α and IL-12 as well as nitric oxide (NO) in response to LPS. Importantly, LPS-tolerance was impaired in SOCS1$^{-/-}$ mice and SOCS1$^{-/-}$ macrophages. Overexpression of SOCS1 in macrophage cell lines results in the suppression of LPS signaling, indicating that SOCS1 negatively regulates not only the JAK/STAT pathway, but also the TLR-NF-kB pathway. We also found that SOCS3 suppresses LPS-sensitivity in mice and macrophages. In this case, IL-10 induced SOCS3, which probably inhibits the MyD88-dependent pathway. Thus, SOCS3 is a long-sought effector molecule for anti-inflammatory effect of IL-10 (Nakagawa et al. 2002).

Clinical application of modulation of intracellular cytokine signaling by SOCS

TNF-α has been most extensively investigated as a target in the efforts to treat RA and IBD. Anti-TNF-α mAbs markedly ameliorate joint involvement in the majority of patients with RA (Berlato et al. 2002; Elliott et al. 1994a). Administration of TNF-α antibodies to CD patients is also shown to be effective. Recently, anti-IL-6 receptor antibodies have been successfully used as a therapy for RA patients (Elliott et al. 1994b). Anti-IL-6 receptor antibodies were also demonstrated to ameliorate T cell-mediated IBD models in mice (Wendling et al. 1993; Yoshizaki et al. 1998).

We recently provided evidence for abnormal cytokine signaling in an animal model of inflammatory synovitis and reported on the utility of forced expression of SOCS3 by adenovirus gene transfer in ameliorating disease (Shouda et al. 2001). We found that SOCS3 transcripts are abundantly expressed in the synovial samples from RA patients, and we noted that RA-derived synoviocytes transfected with a dominant negative form of STAT3 neither proliferated nor secreted IL-6 in response to serum, suggesting that STAT3 is required for the activation of synovial fibroblasts. In agreement with these results, we observed that forced expression of SOCS3 inhibited both synovial fibroblast proliferation and IL-6 production (Shouda et al. 2001).

Supported by these observations, we attempted to express ectopically either SOCS3 or a dominant negative form of STAT3 (dnSTAT3) in two animal models of arthritis to suppress the induction of arthritis (Shouda et al. 2001). We injected an adenovirus construct containing either SOCS3 or dnSTAT3 into the joints of mice susceptible to antigen-induced arthritis and found that joint destruction was prevented in animals expressing either transgene. In the collagen-induced arthritis model, however, the SOCS-3 construct was at all time points more effective than the dnSTAT3 virus in preventing joint damage, probably because SOCS3 can suppress not only STAT3 but also the ras/ERK pathway. In animals with established inflammatory synovitis, gene transfer of SOCS-3 was still helpful in preventing the progression of joint damage.

Our study reinforces the idea that cytokines operating through gp130 are likely important in activating RA synovial fibroblasts. Modulation of the gp130/JAK/STAT pathway therefore represents a reasonable strategy for new anti-inflammatory drug development. Specific JAK kinase inhibitors may have a therapeutic role in treating this and other disorders of the immune system, especially if their toxicity does not preclude their use.

Conclusion

The signal transduction mechanisms of pro- and anti-inflammatory cytokines have recently been uncovered. We have also started to understand how cells regulate these cytokine signal transduction pathways. Especially the negative-feedback circuit of cytokine signaling has been clarified (Fig. 2). Evidence accumulated for the balance of positive and negative pathways is important for the development of inflammation. New therapeutic strategies will emerge from this new knowledge.

References

Akira S, Takeda K, Kaisho T (2001) Toll-like receptors: critical proteins linking innate and acquired immunity. Nat Immunol 2:675–680

Alexander WS, Starr R, Fenner JE, Scott CL, Handman E, Sprigg NS, Corbin JE, Cornish AL, Darwiche R, Owczarek CM, Kay TW, Nicola NA, Hertzog PJ, Metcalf D, Hilton DJ (1999) SOCS1 is a critical inhibitor of interferon γ signaling and prevents the potentially fatal neonatal actions of this cytokine. Cell 98:597–608

Aman MJ, Migone TS, Sasaki A, Ascherman DP, Zhu M, Soldaini E, Imada K, Miyajima A, Yoshimura A, Leonard WJ (1999) CIS associates with the interleukin-2 receptor β chain and inhibits interleukin-2-dependent signaling. J Biol Chem 274:30266–30272

Aso T, Lane WS, Conaway JW, Conaway RC (1995) Elongin (SIII): a multisubunit regulator of elongation by RNA polymerase II. Science 269:1439–1443

Atreya R, Mudter J, Finotto S, Mullberg J, Jostock T, Wirtz S, Schutz M, Bartsch B, Holtmann M, Becker C, Strand D, Czaja J, Schlaak JF, Lehr HA, Autschbach F, Schurmann G, Nishimoto N, Yoshizaki K, Ito H, Kishimoto T, Galle PR, Rose-John S, Neurath MF (2000) Blockade of interleukin 6 trans signaling suppresses T-cell resistance against apoptosis in chronic intestinal inflammation: evidence in Crohn's disease and experimental colitis in vivo. Nat Med 6:583–588

Atsumi T, Ishihara K, Kamimura D, Ikushima H, Ohtani T, Hirota S, Kobayashi H, Park SJ, Saeki Y, Kitamura Y, Hirano T (2002) A point mutation of Tyr-759 in interleukin 6 family cytokine receptor subunit gp130 causes autoimmune arthritis. J Exp Med 196:979–990

Banerjee A, Banks AS, Nawijn MC, Chen XP, Rothman PB (2002) Cutting edge: Suppressor of cytokine signaling 3 inhibits activation of NFATp. J Immunol 168:4277–4281

Berlato C, Cassatella MA, Kinjyo I, Gatto L, Yoshimura A, Bazzoni F (2002) Involvement of suppressor of cytokine signaling-3 as a mediator of the inhibitory effects of IL-10 on lipopolysaccharide-induced macrophage activation. J Immunol 168:6404–6411

Bjorbak C, Lavery HJ, Bates SH, Olson RK, Davis SM, Flier JS, Myers MG Jr (2000) SOCS3 mediates feedback inhibition of the leptin receptor via Tyr985. J Biol Chem 275:40649–40657

Bode JG, Nimmesgern A, Schmitz J, Schaper F, Schmitt M, Frisch W, Haussinger D, Heinrich PC, Graeve L (1999) LPS and TNF-α induce SOCS3 mRNA and inhibit IL-6-induced activation of STAT3 in macrophages .999 LPS and TNF-α induce SOCS3 mRNA and inhibit IL-6-induced activation of STAT3 in macrophages. FEBS Lett 463:365–370

Boisclair YR, Wang J, Shi J, Hurst KR, Ooi GT (2000) Role of the suppressor of cytokine signaling-3 in mediating the inhibitory effects of interleukin-1β on the growth hormone-dependent transcription of the acid-labile subunit gene in liver cells. J Biol Chem 275:3841–3847

Bradsher JN, Jackson KW, Conaway RC, Conaway JW (1993) RNA polymerase II transcription factor SIII: I. Identification, purification, and properties. J Biol Chem 268:25587–25593

Cassatella MA, Gasperini S, Bovolenta C, Calzetti F, Vollebregt M, Scapini P, Marchi M, Suzuki R, Suzuki A, Yoshimura A (1999) Interleukin-10 (IL-10) selectively enhances CIS3/SOCS3 mRNA expression in human neutrophils: evidence for an IL-10-induced pathway that is independent of STAT protein activation. Blood 94:2880–2889

Crepso A, Filla MB, Russell SW, Murphy WJ (2000) Indirect induction of suppressor of cytokine signaling-1 in macrophages stimulated with bacterial lipopolysaccharide: partial role of autocrine/paracrine interferon-a/b. Biochem J 349:99–104

Crespo A, Filla M, Murphy W (2002) Low responsiveness to IFN-γ, after pretreatment of mouse macrophages with lipopolysaccharides, develops via diverse regulatory pathways. Eur J Immunol 32:710–719

Dalpke AH, Opper S, Zimmermann S, Heeg K (2001) Suppressors of cytokine signaling (SOCS)-1 and SOCS-3 are induced by CpG-DNA and modulate cytokine responses in APCs. J Immunol 166:7082–7089

Darnell JE Jr (1997) STATs and gene regulation. Science 277:1630–1635.

De Sepulveda P, Ilangumaran S, Rottapel R (2000) Suppressor of cytokine signaling-1 inhibits VAV function through protein degradation. J. Biol. Chem 275:14005–14008

Diehl S, Anguita J, Hoffmeyer A, Zapton T, Ihle JN, Fikrig E, Rincon M (2000) Inhibition of TH1 differentiation by IL-6 is mediated by SOCS1. Immunity 13:805–815

Duan DR, Pause A, Burgess WH, Aso T, Chen DY, Garrett KP, Conaway RC, Conaway JW, Linehan WM, Klausner RD (1995) Inhibition of transcription elongation by the VHL tumor suppressor protein. Science 269:1402–1406

Duhe RJ, Wang LH, Farrar WL (2001) Negative regulation of Janus kinases. Cell Biochem Biophys 34:17–59

Egwuagu CE, Yu CR, Zhang M, Mahdi RM, Kim SJ, Gery I (2002) Suppressors of cytokine signaling proteins are differentially expressed in Th1 and Th2 cells: implications for Th cell lineage commitment and maintenance. J Immunol 168:3181–3187

Elliott MJ, Maini RN, Feldmann M, Kalden JR, Antoni C, Smolen JS, Leeb B, Breedveld FC, Macfarlane JD, Bijl H, et al (1994a) Randomised double-blind comparison of chimeric monoclonal antibody to tumor necrosis factor α (cA2) versus placebo in rheumatoid arthritis. Lancet 344:1105–1110

Elliott MJ, Maini RN, Feldmann M, Long-Fox A, Charles P, Bijl H, Woody JN (1994b) Repeated therapy with monoclonal antibody to tumor necrosis factor α (cA2) in patients with rheumatoid arthritis. Lancet 344:1125–1127

Endo TA, Masuhara M, Yokouchi M, Suzuki R, Sakamoto H, Mitsui K, Matsumoto A, Tanimura S, Ohtsubo M, Misawa H, Miyazaki T, Leonor N, Taniguchi T, Fujita T, Kanakura Y, Komiya S, Yoshimura A (1997) A new protein containing an SH2 domain that inhibits JAK kinases. Nature 387:921–924

Endo T, Sasaki S, Minoguchi M, Joo A, Yoshimura, A (2003) CIS1 interacts with the Y532 of the prolactin receptor and suppresses prolactin-dependent STAT5 activation. J Biochem (in press)

Ernst M, Inglese M, Waring P, Campbell IK, Bao S, Clay FJ, Alexander WS, Wicks IP, Tarlinton DM, Novak U, Heath JK, Dunn AR (2001) Defective gp130-mediated signal transducer and activator of transcription (STAT) signaling results in degenerative joint disease, gastrointestinal ulceration, and failure of uterine implantation. J Exp Med 194:189–203

Eyles JL, Metcalf D, Grusby MJ, Hilton DJ, Starr R (2002) Negative regulation of interleukin-12 signaling by suppressor of cytokine signaling-1. J Biol Chem 277:43735–43740

Frantsve J, Schwaller J, Sternberg DW, Kutok J, Gilliland, DG (2001) Socs-1 inhibits TEL-JAK2-mediated transformation of hematopoietic cells through inhibition of JAK2 kinase activity and induction of proteasome-mediated degradation. Mol Cell Biol 21:3547–3557

Gearing DP, King JA, Gough NM, Nicola NA (1989) Expression cloning of a receptor for human granulocyte-macrophage colony-stimulating factor. EMBO J 8:3667–3676

Hanada T, Yoshida T, Kinjyo I, Minoguchi S, Yasukawa H, Kato S, Mimata H, Nomura Y, Seki Y, Kubo M, Yoshimura A (2001) A mutant form of JAB/SOCS1 augments the cytokine-induced JAK/STAT pathway by accelerating degradation of wild-type JAB/CIS family proteins through the SOCS-box. J Biol Chem 276:40746–40754

Hanada T, Yoshimura A (2002) Regulation of cytokine signaling and inflammation Cytokine Growth Factor Rev 13:413–421

Hansen JA, Lindberg K, Hilton DJ, Nielsen JH, Billestrup N (1999) Mechanism of inhibition of growth hormone receptor signaling by suppressor of cytokine signaling proteins. Mol Endocrinol 13:1832–1843

Hilton DJ, Richardson RT, Alexander WS, Viney EM, Willson TA, Sprigg NS, Starr R, Nicholson SE, Metcalf D, Nicola NA (1998) Twenty proteins containing a C-terminal SOCS box form five structural classes. Proc Natl Acad Sci USA 95:114–119

Hirano T, Akira S, Taga T, Kishimoto T (1990) Biological and clinical aspects of interleukin 6. Immunol Today 11:443–449

Hong F, Jaruga B, Kim WH, Radaeva S, El-Assal ON, Tian Z, Nguyen VA, Gao B (2002) Opposing roles of STAT1 and STAT3 in T cell-mediated hepatitis: regulation by SOCS. J Clin Invest 110:1503–1513

Hortner M, Nielsch U, Mayr LM, Heinrich PC, Haan S (2002) A new high affinity binding site for suppressor of cytokine signaling-3 on the erythropoietin receptor. Eur J Biochem 269:2516–2526

Ihle JN (1995) Cytokine receptor signaling. Nature 377:591–594

Ihle JN (1996) STATs: signal transducers and activators of transcription. Cell 84:331–334

Ivashkiv LB (2000) Jak-STAT signaling pathways in cells of the immune system. Rev Immunogenet 2:220–230

Jackson PK (2001) A new RING for SUMO: wrestling transcriptional responses into nuclear bodies with PIAS family E3 SUMO ligases. Genes Dev 15:3053–3058

Kamizono S, Hanada T, Yasukawa H, Minoguchi S, Kato R, Minoguchi M, Hattori K, Hatakeyama S, Yada M, Morita S, Kitamura T, Kato H, Nakayama Ki, Yoshimura A (2001) The SOCS box of SOCS-1 accelerates ubiquitin-dependent proteolysis of TEL-JAK2. J Biol Chem 276:12530–12538

Kamura T, Sato S, Haque D, Liu L, Kaelin WG Jr, Conaway RC, Conaway JW (1998) The Elongin BC complex interacts with the conserved SOCS-box motif present in members of the SOCS, ras, WD-40 repeat, and ankyrin repeat families. Genes Dev 12:3872–3881

Kaplan MH, Sun YL, Hoey T, Grusby MJ (1996) Impaired IL-12 responses and enhanced development of Th2 cells in Stat4-deficient mice. Nature 382:174–177

Kibel A, Iliopoulos O, DeCaprio JA, Kaelin WG Jr (1995) Binding of the von Hippel-Lindau tumor suppressor protein to Elongin B and C. Science 269:1444–1446

Kinjyo I, Hanada T, Inagaki-Ohara K, Mori H, Aki D, Ohishi M, Yoshida H, Kubo M, Yoshimura A (2002) SOCS1/JAB Is a negative regulator of LPS-induced macrophage activation. Immunity 17:583–591

Krebs DL, Hilton DJ (2001) SOCS proteins: negative regulators of cytokine signaling. Stem Cells 19:378–387

Liu B, Gross M, ten Hoeve J, Shuai K (2001) A transcriptional corepressor of Stat1 with an essential LXXLL signature motif. Proc Natl Acad Sci USA 98:3203–3207

Marine JC, McKay C, Wang D, Topham DJ, Parganas E, Nakajima H, Pendeville H, Yasukawa H, Sasaki A, Yoshimura A, Ihle JN (1999) SOCS3 is essential in the regulation of fetal liver erythropoiesis. Cell 98:617–627

Marine JC, Topham DJ, McKay C, Wang D, Parganas E, Stravopodis D, Yoshimura A, Ihle JN (1999) SOCS1 deficiency causes a lymphocyte-dependent perinatal lethality. Cell 98:609–616

Masuhara M, Sakamoto H, Matsumoto A, Suzuki R, Yasukawa H, Mitsui K, Wakioka T, Tanimura S, Sasaki A, Misawa H, Yokouchi M, Ohtsubo M, Yoshimura A (1997) Cloning and characterization of novel CIS family genes. Biochem Biophys Res Commun 239:439–446

Matsuda T, Yamamoto T, Kishi H, Yoshimura A, Muraguchi A (2000) SOCS-1 can suppress CD3- and Syk-mediated NF-AT activation in a nonlymphoid cell line. FEBS Lett 472:235–240

Matsumoto A, Seki Y, Kubo M, Ohtsuka S, Suzuki A, Hayashi I, Tsuji K, Nakahata T, Okabe M, Yamada S, Yoshimura A (1999) Suppression of STAT5 functions in liver, mammary glands, and T cells in cytokine-inducible SH2-containing protein 1 transgenic mice. Mol Cell Biol 19:6396–6407

McHugh RS, Whitters MJ, Piccirillo CA, Young DA, Shevach EM, Collins M, Byrne MC (2002) CD4+CD25+ Immunoregulatory T Cells: gene expression analysis reveals a functional role for the glucocorticoid-induced TNF-α receptor. Immunity 16:311–323

Metcalf D, Di Rago L, Mifsud S, Hartley L, Alexander WS (2000) The development of fatal myocarditis and polymyositis in mice heterozygous for IFN-γ and lacking the SOCS-1 gene. Proc Natl Acad Sci USA 97:9174–9179

Metcalf D, Mifsud S, Di Rago L, Nicola NA, Hilton DJ, Alexander WS (2002) Polycystic kidneys and chronic inflammatory lesions are the delayed consequences of loss of the suppressor of cytokine signaling-1 (SOCS-1) Proc Natl Acad Sci USA 99:943–948

Morita Y, Naka T, Kawazoe Y, Fujimoto M, Narazaki M, Nakagawa R, Fukuyama H, Nagata S, Kishimoto T (2000) Signals transducers and activators of transcription (STAT)-induced STAT inhibitor-1 (SSI-1)/

suppressor of cytokine signaling-1 (SOCS-1) suppresses tumor-necrosis-factor-a-induced cell death in fibroblasts. Proc Natl Acad Sci USA 97:5405–5410

Nagai H, Kim Y, Konishi N, Baba M, Kubota T, Yoshimura A, Emi M (2002) Combined hypermethylation and chromosome loss associated with inactivation of SSI-1/SOCS-1/JAB gene in human hepatocellular carcinomas. Cancer Lett 186:59–64

Naka T, Matsumoto T, Narazaki M, Fujimoto M, Morita Y, Ohsawa Y, Saito H, Nagasawa T, Uchiyama Y, Kishimoto T (1998) Accelerated apoptosis of lymphocytes by augmented induction of Bax in SSI-1 (STAT-induced STAT inhibitor-1) deficient mice. Proc Natl Acad Sci USA 95:15577–1582

Naka T, Tsutsui H, Fujimoto M, Kawazoe Y, Kohzaki H, Morita Y, Nakagawa R, Narazaki M, Adachi K, Yoshimoto T, Nakanishi K, Kishimoto T (2001) SOCS-1/SSI-1-deficient NKT cells participate in severe hepatitis through dysregulated cross-talk inhibition of IFN-γ and IL-4 signaling in vivo. Immunity 14:535–545

Naka T, Narazaki M, Hirata M, Matsumoto T, Minamoto S, Aono A, Nishimoto N, Kajita T, Taga T, Yoshizaki K, Akira S, Kishimoto T (1997) Structure and function of a new STAT-induced STAT inhibitor. Nature 387:924–929

Nakagawa R, Naka T, Tsutsui H, Fujimoto M, Kimura A, Abe T, Seki E, Sato S, Takeuchi O, Takeda K, Akira S, Yamanishi K, Kawase I, Nakanishi K, Kishimoto T (2002) SOCS-1 participates in negative regulation of LPS responses. Immunity 17:677–687

Nicholson SE, De Souza D, Fabri LJ, Corbin J, Willson TA, Zhang JG, Silva A, Asimakis M, Farley A, Nash AD, Metcalf D, Hilton DJ, Nicola NA, Baca M (2000) Suppressor of cytokine signaling-3 preferentially binds to the SHP-2-binding site on the shared cytokine receptor subunit gp130. Proc Natl Acad Sci USA 97:6493–6498

O'Farrell AM, Liu Y, Moore KW, Mui AL (1998) IL-10 inhibits macrophage activation and proliferation by distinct signaling mechanisms: evidence for Stat3-dependent and -independent pathways. EMBO J 17:1006–1018

Parrello T, Monteleone G, Cucchiara S, Monteleone I, Sebkova L, Doldo P, Luzza F, Pallone F (2000) Upregulation of the IL-12 receptor β 2 chain in Crohn's disease. J Immunol 165:7234–7239

Pezet A, Favre H, Kelly PA, Edery M (1999) Inhibition and restoration of prolactin signal transduction by suppressors of cytokine signaling. J Biol Chem 274:24497–24502

Roberts AW, Robb L, Rakar S, Hartley L, Cluse L, Nicola NA, Metcalf D, Hilton DJ, Alexander WS (2001) Placental defects and embryonic lethality in mice lacking suppressor of cytokine signaling 3. Proc Natl Acad Sci USA 98:9324–9329

Rui L, Yuan M, Frantz D, Shoelson S, White MF (2002) SOCS-1 and SOCS-3 block insulin signaling by ubiquitin-mediated degradation of IRS1 and IRS2 J Biol Chem 277:42394–42398

Sampath D, Castro M, Look DC, Holtzman MJ (1999) Constitutive activation of an epithelial signal transducer and activator of transcription (STAT) pathway in asthma. J Clin Invest 103:1353–1361

Sasaki A, Yasukawa H, Shouda T, Kitamura T, Dikic I, Yoshimura A (2000) CIS3/SOCS3 suppresses erythropoietin signaling by binding the EPO receptor and JAK2. J Biol Chem 275:29338–29347

Sasaki A, Yasukawa H, Suzuki A, Kamizono S, Syoda T, Kinjyo I, Sasaki M, Johnston JA, Yoshimura A (1999) Cytokine-inducible SH2 protein-3 (CIS3/SOCS3) inhibits Janus tyrosine kinase by binding through the N-terminal kinase inhibitory region as well as SH2 domain. Genes Cells 4:339–351

Schmitz J, Weissenbach M, Haan S, Heinrich P C, Schaper F (2000) SOCS3 exerts its inhibitory function on interleukin-6 signal transduction through the SHP2 recruitment site of gp130. J Biol Chem 275:12848–12856

Seki Y, Hayashi K, Matsumoto A, Seki N, Tsukada J, Ransom J, Naka T, Kishimoto T, Yoshimura A, Kubo M (2002) Expression of the suppressor of cytokine signaling-5 (SOCS5) negatively regulates IL-4-dependent STAT6 activation and Th2 differentiation. Proc Natl Acad Sci USA 99:13003–13008

Shimoda K, van Deursen J, Sangster MY, Sarawar SR, Carson RT, Tripp RA, Chu C, Quelle FW, Nosaka T, Vignali DA, Doherty PC, Grosveld G, Paul WE, Ihle JN (1996) Lack of IL-4-induced Th2 response and IgE class switching in mice with disrupted Stat6 gene. Nature 380:630–633

Shirogane T, Fukada T, Muller JM, Shima DT, Hibi M, Hirano T (1999) Synergistic roles for Pim-1 and c-Myc in STAT3-mediated cell cycle progression and antiapoptosis. Immunity 11:709–719

Shouda T, Yoshida T, Hanada T, Wakioka T, Oishi M, Miyoshi K, Komiya S, Kosai K, Hanakawa Y, Hashimoto K, Nagata K, Yoshimura A (2001) Induction of the cytokine signal regulator SOCS3/CIS3 as a therapeutic strategy for treating inflammatory arthritis. J Clin Invest 108:1781–1788

Starr R, Metcalf D, Elefanty AG, Brysha M, Willson TA, Nicola NA, Hilton DJ, Alexander WS (1998) Liver degeneration and lymphoid deficiencies in mice lacking suppressor of cytokine signaling-1. Proc Natl Acad Sci USA 95:14395–14399

Starr R, Willson TA, Viney EM, Murray LJ, Rayner JR, Jenkins BJ, Gonda TJ, Alexander WS, Metcalf D, Nicola NA, Hilton DJ (1997) A family of cytokine-inducible inhibitors of signalling. Nature 387:917–921

Stoiber D, Kovarik P, Cohney S, Johnston JA, Steinlein P, Decker T (1999) Lipopolysaccharide induces in macrophages the synthesis of the suppressor of cytokine signaling 3 and suppresses signal transduction in response to the activating factor IFN-γ. J Immunol 163:2640–2647

Suzuki A, Hanada T, Mitsuyama K, Yoshida T, Kamizono S, Hoshino T, Kubo M, Yamashita A, Okabe M, Takeda K, Akira S, Matsumoto S, Toyonaga A, Sata M, Yoshimura A (2001) CIS3/SOCS3/SSI3 plays a negative regulatory role in STAT3 activation and intestinal inflammation. J Exp Med 193:471–481

Takeda K, Clausen BE, Kaisho T, Tsujimura T, Terada N, Forster I, Akira S (1999) Enhanced Th1 activity and development of chronic enterocolitis in mice devoid of Stat3 in macrophages and neutrophils. Immunity 10:39–49

Takeda K, Kaisho T, Yoshida N, Takeda J, Kishimoto T Akira S (1998) Stat3 activation is responsible for IL-6-dependent T-cell proliferation through preventing apoptosis: generation and characterization of T cell-specific Stat3-deficient mice. J Immunol 161:4652–4660

Takeda K, Tanaka T, Shi W, Matsumoto M, Minami M, Kashiwamura S, Nakanishi K, Yoshida N, Kishimoto T, Akira S (1996) Essential role of Stat6 in IL-4 signaling. Nature 380:627–630

Tebbutt NC, Giraud AS, Inglese M, Jenkins B, Waring P, Clay FJ, Malki S, Alderman BM, Grail D, Hollande F, Heath JK, Ernst M (2002) Reciprocal regulation of gastrointestinal homeostasis by SHP2 and STAT-mediated trefoil gene activation in gp130 mutant mice. Nat Med 8:1089–1097

Tonko-Geymayer S, Goupille O, Tonko M, Soratroi C, Yoshimura A, Streuli C, Ziemiecki A, Kofler R, Doppler W (2002) Regulation and function of the cytokine-inducible SH-2 domain proteins, CIS and SOCS3, in mammary epithelial cells. Mol Endocrinol 16:1680–1695

Ungureanu D, Saharinen P, Junttila I, Hilton DJ, Silvennoinen O (2002) Regulation of Jak2 through the ubiquitin-proteasome pathway involves phosphorylation of Jak2 on Y1007 and interaction with SOCS-1. Mol Cell Biol 22:3316–3326

Verdier F, Chretien S, Muller O, Varlet P, Yoshimura A, Gisselbrecht S, Lacombe C, Mayeux P (1988) Proteasomes regulate erythropoietin receptor and signal transducer and activator of transcription 5 (STAT5) activation. Possible involvement of the ubiquitinated Cis protein. J Biol Chem 273:28185–28190

Wendling D, Racadot E, Wijdenes J (1993) Treatment of severe rheumatoid arthritis by antiinterleukin 6 monoclonal antibody. J Rheumatol 20:259–262

Wirtz S, Finotto S, Kanzler S, Lohse AW, Blessing M, Lehr HA, Galle PR, Neurath MF (1999) Cutting edge: chronic intestinal inflammation in STAT-4 transgenic mice: characterization of disease and adoptive transfer by TNF-α plus IFN-γ-producing CD4+ T cells that respond to bacterial antigens. J Immunol 162:1884–1888

Yamamoto M, Yoshizaki K, Kishimoto T, Ito H (2000) IL-6 is required for the development of Th1 cell-mediated murine colitis. J Immunol 164:4878–4882

Yasukawa H, Misawa H, Sakamoto H, Masuhara M, Sasaki A, Wakioka T, Ohtsuka S, Imaizumi T, Matsuda T, Ihle JN, Yoshimura A (1999) The JAK-binding protein JAB inhibits Janus tyrosine kinase activity through binding in the activation loop. EMBO J 18:1309–1320

Yasukawa H, Sasaki A, Yoshimura A (2000) Negative regulation of cytokine signaling pathways. Annu Rev Immunol 18:143–164

Yoshikawa H, Matsubara K, Qian GS, Jackson P, Groopman JD, Manning JE, Harris CC, Herman JG (2001) SOCS-1, a negative regulator of the JAK/STAT pathway, is silenced by methylation in human hepatocellular carcinoma and shows growth-suppression activity. Nat Genet 28:29–35

Yoshimura A, Ichihara M, Kinjyo I, Moriyama M, Copeland NG, Gilbert DJ, Jenkins NA, Hara T, Miyajima A (1995) A novel cytokine-inducible gene CIS encodes an SH2 containing protein that binds to tyrosine-phosphorylated interleukin 3 and erythropoietin receptors. EMBO J 14:2816–2826

Yoshizaki K, Nishimoto N, Mihara M, Kishimoto T (1998) Therapy of rheumatoid arthritis by blocking IL-6 signal transduction with a humanized anti-IL-6 receptor antibody. Springer Semin Immunopathol 20:247–259

Yu CL, Burakoff SJ (1997) Involvement of proteasomes in regulating Jak-STAT pathways upon interleukin-2 stimulation. J Biol Chem 272:14017–14020

Zhang JG, Farley A, Nicholson SE, Willson TA, Zugaro LM, Simpson RJ, Moritz RL, Cary D, Richardson R, Hausmann G, Kile BJ, Kent SB, Alexander WS, Metcalf D, Hilton DJ, Nicola NA, Baca M (1999) The conserved SOCS box motif in suppressors of cytokine signaling binds to elongins B and C and may couple bound proteins to proteasomal degradation. Proc Natl Acad Sci USA 96:2071–2076

Zhang JG, Metcalf D, Rakar S, Asimakis M, Greenhalgh CJ, Willson TA, Starr R, Nicholson SE, Carter W, Alexander WS, Hilton DJ, Nicola NA (2001) The SOCS box of suppressor of cytokine signaling-1 is important for inhibition of cytokine action in vivo. Proc Natl Acad Sci USA 98:13261–13265

J. Kalesnikoff · L. M. Sly · M. R. Hughes · T. Büchse · M. J. Rauh · L.-P. Cao ·
V. Lam · A. Mui · M. Huber · G. Krystal

The role of SHIP in cytokine-induced signaling

Published online: 12 April 2003
© Springer-Verlag 2003

Abstract The phosphatidylinositol (PI)-3 kinase (PI3K) pathway plays a central role in regulating many biological processes via the generation of the key second messenger PI-3,4,5-trisphosphate (PI-3,4,5-P_3). This membrane-associated phospholipid, which is rapidly, albeit transiently, synthesized from PI-4,5-P_2 by PI3K in response to a diverse array of extracellular stimuli, attracts pleckstrin homology (PH) domain-containing proteins to membranes to mediate its many effects. To ensure that the activation of this pathway is appropriately suppressed/terminated, the ubiquitously expressed tumor suppressor PTEN hydrolyzes PI-3,4,5-P_3 back to PI-4,5-P_2 while the 145-kDa hemopoietic-restricted *SH2*-containing *i*nositol 5′-*p*hosphatase, SHIP (also known as SHIP1), the 104-kDa *s*tem cell-restricted SHIP (sSHIP) and the more widely expressed 150-kDa SHIP2 hydrolyze PI-3,4,5-P_3 to PI-3,4-P_2. In this review we will concentrate on the properties of the three SHIPs, with special emphasis being placed on the role that SHIP plays in cytokine-induced signaling.

Abbreviations *BCR:* B cell receptor · *BMMCs:* Bone marrow derived mast cells ·
Epo: Erythropoietin · *ES Cells:* embryonic stem cells · *GM-CSF:* Granulocyte macrophage colony stimulating factor · *IL-3:* Interleukin-3 · *IP$_4$:* Inositol-1,3,4,5-tetrakisphosphate · *M-CSF:* Macrophage colony stimulating factor ·
PH: Pleckstrin homology · *PI3K:* Phosphatidylinositol-3 kinase · *PI-3,4,5-P_3:* Phosphatidylinositol-3,4,5-trisphosphate · *SHIP:* Src homology 2 containing inositol

---

J. Kalesnikoff · L. M. Sly · M. R. Hughes · T. Büchse · M. J. Rauh · L.-P. Cao · V. Lam ·
G. Krystal (✉)
The Terry Fox Laboratory, B.C. Cancer Agency, Vancouver, V5Z 1L3, Canada
e-mail: gkrystal@bccancer.bc.ca

A. Mui
The Jack Bell Research Centre, Vancouver Hospital and Health Sciences Centre,
Vancouver, Canada

M. Huber
Department of Molecular Immunology, Biology III,
University of Freiburg and Max-Planck-Institute for Immunobiology,
79108 Freiburg, Germany

5′-phosphatase · *SF:* Steel Factor · *sSHIP:* Stem cell SHIP · *TCR:* T cell receptor · *TPO:* Thrombopoietin · *WT:* Wild type

Introduction

It is now well established that the phosphatidylinositol (PI)-3 kinase (PI3K) pathway plays a central role in regulating many cellular decisions. These include, depending on the cell type, survival, adhesion, movement, proliferation, differentiation, and end cell activation (Krystal 2000). A key second messenger in this pathway is the membrane-associated PI-3,4,5-trisphosphate (PI-3,4,5-P_3), which is present at low levels in unstimulated cells but is rapidly synthesized from PI-4,5-P_2 by PI3K in response to a diverse array of extracellular stimuli. This transiently generated PI-3,4,5-P_3 attracts pleckstrin homology (PH) domain-containing proteins to the plasma membrane to mediate its effects (Rameh and Cantley 1999; Huber et al. 1999). To ensure that the activation of this pathway is appropriately suppressed/terminated, the ubiquitously expressed tumor suppressor PTEN hydrolyzes PI-3,4,5-P_3 back to PI-4,5-P_2 (Maehama and Dixon 1998; Stambolic et al. 1998) while the 145-kDa hemopoietic-restricted *SH2*-containing inositol 5′-*p*hosphatase, SHIP (also known as SHIP1; Huber et al. 1999), the 104-kDa *s*tem cell-restricted SHIP (sSHIP; Tu et al. 2001) and the more widely expressed 150-kDa SHIP2 (Pesesse et al. 1997; Wisniewski et al. 1999; Pesesse et al. 1998; Muraille et al. 1999) break it down to PI-3,4-P_2 (Fig. 1A). The fact that almost 50% of human cancers contain biallelic inactivating mutations of PTEN (Cantley and Neel 1999) highlights the importance of these phospholipid phosphatases in preventing uncontrolled cell growth. In this review we concentrate on the properties of the three SHIPs, with special emphasis on the role that SHIP plays in cytokine-induced signaling.

The properties of SHIP, sSHIP, and SHIP2

In 1996, we (Damen et al., Lioubin et al., and Kavanaugh et al.) independently cloned the cDNA of a 145-kDa intracellular protein that became both tyrosine phosphorylated and associated with the adaptor protein, Shc, after cytokine, growth factor, or immunoreceptor stimulation of hemopoietic cells (Liu et al. 1994). Its predicted amino acid sequence revealed an amino-terminal SH2 domain that binds preferentially to the sequence pY(Y/D)X(L/I/V) (Osborne et al. 1996), a centrally located 5′-phosphatase domain that selectively hydrolyzes PI-3,4,5-P_3, and inositol-1,3,4,5-tetrakisphosphate (IP_4) in vitro (Damen et al. 1996) and in vivo (Huber et al. 1999a; Valderrama-Carvajal et al. 2002), two NPXY sequences that, when phosphorylated, bind the phosphotyrosine binding (PTB) domains of Shc (Huber et al. 1999), Dok1 (Sattler et al. 2001), and Dok2 (Tamir et al. 2000), and a critical proline rich C-terminus that binds a subset of SH3-containing proteins (Wisniewski et al. 1999).

During murine development, SHIP is first detectable, by RT-PCR, in 7.5-day postcoitus mouse embryos (coincident with and dependent upon the onset of hemopoiesis) and its protein expression pattern in the embryo appears restricted to hemopoietic cells (Liu et al. 1998). In the adult mouse, SHIP protein expression is also restricted to hemopoietic cells (and to spermatids; Liu et al. 1998). Also noteworthy is that SHIP protein expression ap-

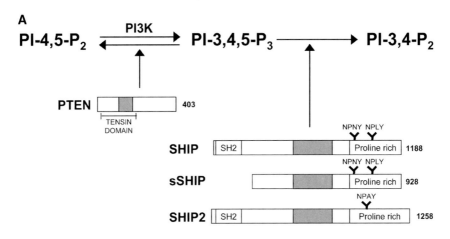

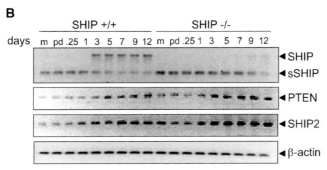

Fig. 1 (**A**) The structures of SHIP, sSHIP, SHIP2 and PTEN. The numbers to the right of the proteins refer to the total number of amino acids. (**B**) Expression of SHIP, sSHIP, PTEN and SHIP2 in SHIP$^{+/+}$ and $-/-$ ES cells. Semiquantitative RT-PCR of lipid phosphatase expression by ES cells differentiating into embryoid bodies (EBs). RNA was isolated from SHIP$^{+/+}$ and $-/-$ ES cells growing in maintenance medium (m), pre-differentiation medium (pd) or during differentiation into EBs by removal of LIF from their medium for increasing lengths of time (shown in days). RT-PCR for each of the lipid phosphatases SHIP, sSHIP, PTEN and SHIP2 were performed at limiting PCR cycles. These were compared to semiquantitative RT-PCR of the housekeeping gene, β-actin, a a control. SHIP$^{-/-}$ ES cells do not express SHIP while SHIP$^{+/+}$ ES cells show SHIP expression 3 days after differentiation coinciding with the onset of hemopoiesis in the EBs (3.5 days). SHIP$^{-/-}$ ES cells have increased levels of sSHIP expression in undifferentiated and differentiating EBs. As well, a higher level of expression of sSHIP is maintained throughout EB formation. SHIP2 and PTEN expression levels are also slightly higher in differentiating SHIP$^{-/-}$ ES cells notably at 3 days after differentiation when WT cells are beginning to express SHIP protein

pears to vary considerably during hemopoiesis (Geier et al. 1997; Liu et al. 1998), increasing substantially, for example, with T cell maturation (Liu et al. 1998) and showing a bimodal expression pattern during B cell development and a dramatic increase when resting B cells are activated (Brauweiler et al. 2001). Complicating the situation, two alternate splice forms of SHIP have been discovered and their expression levels also change during hemopoiesis (Lucas and Rohrschneider 1999; Wolf et al. 2000). In addition, C-terminal truncations of SHIP have been identified in nonionic detergent solubilized cell lysates that may (Damen et al. 1998) or may not (Horn et al. 2001) exist in vivo.

In addition to SHIP, there is a 104-kDa sSHIP that is only expressed in murine embryonic stem (ES) cells and hemopoietic stem cells (Tu et al. 2001). This sSHIP, which is the murine homolog of the human SIP-110 cloned by Kavanaugh et al. in 1996, lacks the SH2 domain of full-length SHIP and is generated by transcription from a promoter within the intron between exons 5 and 6 of the SHIP gene. It is replaced by full-length SHIP as hemopoietic progenitors differentiate and, because it lacks an SH2 domain, it is neither tyrosine phosphorylated nor associated with Shc following stimulation (Tu et al. 2001). However, it does bind constitutively to Grb2 and may be recruited via Grb2's SH2 domain to the plasma membrane in response to extracellular stimuli to regulate PI-3,4,5-P_3 levels in stem cells (Tu et al. 2001).

In addition to SHIP and sSHIP, there is a more widely expressed 150-kDa protein, SHIP2, that is encoded by a separate gene. Its overall structure is similar to SHIP (Fig. 1A) and, like SHIP and sSHIP, specifically hydrolyzes the 5-phosphate from PI-3,4,5-P_3 in vitro (Pesesse et al. 1998) and in vivo (Dyson et al. 2001) and may (Pesesse et al. 1998) or may not (Wisniewski et al. 1999) hydrolyze IP_4 as well. It also becomes tyrosine phosphorylated and associated with Shc in response to extracellular stimuli (Krystal 2000) and exists, like SHIP and sSHIP, in lower molecular weight forms (Wisniewski et al. 1999). Interestingly, SHIP2's proline-rich C-terminus (Pesesse et al. 1997) is very different from that of SHIP and, since we (Damen et al. 2001) and Aman et al. (2000) have shown that SHIP's C-terminus is essential for its translocation to the plasma membrane following stimulation in bone marrow-derived mast cells (BMMCs) and B cells, respectively, this could allow for some nonredundancy in the regulation of SHIP and SHIP2. Consistent with this, Wisniewski et al. (1999) have reported that SHIP binds to the SH3 domains of Grb2 and Src, while SHIP2 binds to the SH3 domain of Abl but not to Grb2. We have found as well that the C-terminal SH3 domain of Grb2, which we used originally to purify SHIP (Damen et al. 1996), does not bind SHIP2 (T. Büchse and G. Krystal, unpublished data). Also worthy of note is that both SHIP and SHIP2 are constitutively tyrosine phosphorylated and associated with Shc in chronic myelogenous leukemia (CML) progenitor cells (Wisniewski et al. 1999) and that SHIP is reduced in both primary cells from leukemic patients and following induced expression of BCR-ABL in Ba/F3 cells (Sattler et al. 1997). In fact, Sattler et al. (1997) have shown that there is an inverse relationship between the expression of BCR-ABL and SHIP, suggesting that reduced SHIP activity might be a prerequisite for the proliferative advantage of some chronic and acute myelogenous leukemic clones. It is thus possible that SHIP and SHIP2 act as tumor suppressors during myelopoiesis.

The phenotype of SHIP-deficient mice

In 1998, we generated a SHIP knockout (–/–) mouse in collaboration with Dr. K. Humphries by replacing SHIP's first exon with the neomycin resistance gene in the antisense orientation (Helgason et al. 1998). Although these mice are viable, they have a shortened lifespan and overproduce granulocytes and macrophages, suffer from progressive splenomegaly, extramedullary hemopoiesis, massive myeloid infiltration of the lungs (Helgason et al. 1998), and osteoporosis (because of an increased number and bone resorbing activity of their osteoclasts; Takeshita et al. 2002). Related to this, a subset of patients with familial Paget-like osteoporosis show a loss of heterozygosity at chromosome 2q36 (Hocking et al. 2001), the chromosomal location of human SHIP (Ware et al. 1996) and this could suggest that the osteoporosis in these patients is due to a reduction in SHIP levels. Our SHIP$^{-/-}$ phenotype, which has been subsequently confirmed by Liu et al. (1998), is reminiscent of normal mice transplanted with bone marrow cells overexpressing granulocyte macrophage colony stimulating factor (GM-CSF; Johnson et al. 1989) and may therefore suggest that the SHIP$^{-/-}$ mouse pathology is due in large part to the hyperresponsiveness of granulocyte/macrophage progenitors. Relevant to this, these SHIP$^{-/-}$ progenitors are substantially more responsive to a number of cytokines and growth factors [e.g., GM-CSF, interleukin-3 (IL-3), macrophage colony stimulating factor (M-CSF) and Steel Factor (SF)] than their SHIP$^{+/+}$ counterparts and, even in the absence of added growth factors, develop into small colonies (Helgason et al. 1998). we found that SHIP$^{-/-}$ progenitors are substantially more responsive to chemokines (Kim et al. 1999) as well. These findings are consistent with SHIP being a negative regulator of myeloid cell proliferation/survival and chemotaxis. SHIP$^{-/-}$ mice share many phenotypic characteristics with PTEN$^{+/-}$ mice (Fox et al. 2002) as well, suggesting that it is the higher levels of PI-3,4,5-P$_3$ in SHIP$^{-/-}$ mice that are primarily responsible for its phenotype.

Interestingly, SHIP2 deficient mice possess a far more severe phenotype than that observed with SHIP$^{-/-}$ mice and die perinatally from insulin hypersensitivity-induced hypoglycemia (Clement et al. 2001). This difference in severity is most likely because SHIP2 is expressed to some degree in hemopoietic cells (Muraille et al. 1999), while SHIP is not expressed in nonhemopoietic tissues (Liu et al. 1998) and therefore cannot compensate for the loss of SHIP2. In addition, as shown in Fig. 1B, sSHIP is still expressed in our SHIP$^{-/-}$ ES cells and its expression is higher and more prolonged than in SHIP$^{+/+}$ ES cells when induced to differentiate (L.M. Sly and G. Krystal, unpublished data). Also, as can be seen in Fig. 1B, SHIP2 and PTEN appear to be slightly upregulated in our SHIP$^{-/-}$ ES cells, at least as assessed by semiquantitative RT-PCR.

The mechanism of action of SHIP

Since SHIP's 5-phosphatase activity does not appear to change with extracellular stimulation (Damen et al. 1996) or tyrosine phosphorylation (Phee et al. 2000), the current consensus is that SHIP mediates its inhibitory effects by translocating to sites of synthesis of PI-3,4,5-P$_3$ (and perhaps IP$_4$; Phee et al. 2000). To understand what regulates this translocation we and others have been searching for SHIP binding partners that might play a role in this process. In support of this approach we have found that lysates from SHIP$^{-/-}$ BMMCs dramatically increase the affinity of recombinant SHIP for the tyrosine-phosphorylated *i*mmunoreceptor *t*yrosine-based *i*nhibition *m*otif (pITIM) of the FcγRIIB in in vitro assays (L.-P. Cao and G. Krystal, unpublished data). What has been found to date is that SHIP associates, via its SH2 and NPXpYs, with the pY$^{317\ \text{or}\ 239}$ (Velazquez et al. 2000) and PTB motifs of Shc, respectively, following stimulation of myeloid (Liu et al. 1997) and B cell (Tridandapani et al. 1999) lines and via SHIP's NPXpYs with Shc's PTB domain in T cell receptor-activated T cells (Lamkin et al. 1997). We (Liu et al. 1997a) and Sattler et al. (1997) have also found that SHIP binds, via its SH2 domain, with the tyrosine phosphatase, SHP-2 (via the latter's pY$^{542\ \text{or}\ 580}$). Moreover, kinetic studies suggest that, following cytokine stimulation, SHIP/Shc complexes form first and are then replaced by SHIP/SHP-2 complexes, and it is possible that SHP-2 is responsible for the subsequent dephosphorylation of SHIP (Liu et al. 1997a). SHIP also has been found to bind to certain adaptor proteins like the Doks (van Dijk et al. 2000; Dunant et al. 2000; Lemay et al. 2000) and Gabs (Koncz et al. 2001; Liu et al. 2001) and this may facilitate the inhibition of Ras-mediated signaling (Ott et al. 2002) and the formation of larger complexes (Koncz et al. 2001; Liu et al. 2001), respectively. We and others have also been exploring what domains within SHIP are required for its translocation and have found that SHIP's SH2 domain and its proline-rich C-terminus are both critical for SHIP's tyrosine phosphorylation, its association with either Shc or SHP-2, its translocation to the plasma membrane, and its biological effects (Liu et al. 1997; Damen et al. 2001; Aman et al. 2000).

To elucidate the role of SHIP's binding partners in translocating SHIP to the plasma membrane, Marchetto et al. (1999), overexpressed Shc and found it increased FLT3-induced SHIP tyrosine phosphorylation in Ba/F3 cells expressing the FLT3R and limited FLT3-dependent cell growth. Since we have strong evidence that the plasma membrane-associated Src family of tyrosine kinases is responsible for phosphorylating SHIP, regardless of the extracellular stimulus (M.D. Ware and G. Krystal, manuscript submitted), this suggests that Shc, because it regulates the tyrosine phosphorylation of SHIP, may be taking SHIP to the plasma membrane (Marchetto et al. 1999). More recently, Tridandapani et al. (2002) reported that during phagocytic FcγR activation in human myeloid cells, overexpressing a dominant negative mutant of Shc (where $Y^{239,240\ and\ 317}$ were replaced with phenylalanines) inhibited SHIP's tyrosine phosphorylation. In addition, Galandrini et al. (2001) showed, by overexpressing either WT Shc or the SH2 domain of Shc (a dominant negative form), that Shc is involved in bringing SHIP to CD16 to reduce CD16-induced antibody-dependent cellular cytotoxicity (ADCC) of NK cells. Taken together, these results suggest that, at least in some cells and in response to some extracellular stimuli, Shc is an important player in getting SHIP to the plasma membrane to hydrolyze PI-3,4,5-P$_3$.

However, in IgE + antigen (Ag)-stimulated BMMCs, we found that SHIP was required for Shc tyrosine phosphorylation, suggesting that in these cells and with this stimulus, Shc cannot get to the plasma membrane without SHIP (Huber et al. 1998a). Consistent with this model, Tridandapani et al. (1999) have good evidence that when the inhibitory FcγRIIB is coclustered with the activated B cell receptor (BCR) in B cells that SHIP, via its SH2 domain, first binds to the pITIM of FcγRIIB. This leads to the phosphorylation of SHIP by Lyn on its NPXYs, which attracts Shc via its PTB domain. Shc in turn gets tyrosine phosphorylated by Lyn, which enables it to compete with the FcγRIIB for SHIP's SH2 domain and wrest SHIP away from the plasma membrane and back into the cytosol.

Adding another layer of complexity, Harmer and DeFranco (1999), using Grb2$^{+/+}$ and $^{-/-}$ B cell lines, have concluded that Grb2, Shc, and SHIP form a ternary complex and that Grb2 stabilizes the Shc/SHIP complex. Although we initially did not observe Grb2 in SHIP/Shc complexes in myeloid cell lines (Liu et al. 1997), we have since found, by immunoprecipitating with a newly available anti-Grb2 antibody (Santa Cruz, cat # sc-255-G) that SHIP does indeed associate constitutively via its C-terminus with Grb2 in these cells. Since, as mentioned earlier, the C-terminus of SHIP is critical for its translocation, this could mean that Grb2 plays a role in this process.

Worthy of note is that, at least in some cells, SHIP has been shown to bind to the cytoskeleton as well. For example, thrombin stimulation of human blood platelets, which express endogenous SHIP, leads to the tyrosine phosphorylation and translocation of SHIP to the actin cytoskeleton, and this correlates with the accumulation of PI-3,4-P$_2$ in these cells (Giuriato et al. 1997). In addition, Dyson et al. (2001) have found, using a yeast two-hybrid screen, that the SHIP-related protein, SHIP2, binds the actin-binding proteins filamin A, B, and C and these filamins may be responsible for SHIP2 localizing to the actin cytoskeleton and regulating PI-3,4,5-P_3 levels there. Similar actin binding proteins may be involved in translocating SHIP to the cytoskeleton.

In terms of what SHIP binds to at the plasma membrane, it has been shown to be recruited via its SH2 domain to certain pITIM-containing inhibitory coreceptors such as the FcγRIIB or MAFA [to inhibit FcεR1-induced degranulation of mast cells (Ono et al. 1996; Ono et al. 1997; Vely et al. 1997; Tridandapani et al. 1997; Xu et al. 2001) and to certain tyrosine-phosphorylated immunoreceptor *t*yrosine based *a*ctivation *m*otif (ITAM)-containing proteins, such as the β (Kimura et al. 1997) and γ (Osborne et al. 1996) subunits of the FcεRI and the ζ chain of the T cell receptor (Osborne et al. 1996). However, we have been unsuccessful in our attempts to demonstrate any in vivo association, via coprecipitation studies with various detergents and buffers, between SHIP and the FcεRI, c-kit, the IL-3R, or the EpoR (M. Huber and G. Krystal, unpublished data). However, Tridandapani et al. (2002) have recently reported that during phagocytic FcγR activation in human myeloid cells, SHIP becomes tyrosine phosphorylated and can be coprecipitated with the native ITAM-bearing FcγRIIa. In addition, SHIP may not always bind directly to an activated cell surface receptor, but rather to a transmembrane protein that becomes tyrosine phosphorylated following activation of a receptor. For example, Mikhalap et al. (1999) have reported that BCR activation leads to the tyrosine phosphorylation of an 80-kDa transmembrane protein, CD150 (also known as SLAM), which is highly expressed in activated B, T, and dendritic cells, and this protein then binds SHIP. This results in SHIP becoming tyrosine phosphorylated, most likely by the Src family members, Fgr and Lyn, which also associate with the tyrosine-phosphorylated form of CD150. Of special interest, the gene responsible for X-linked lymphoproliferative syndrome (XLP), which is characterized by an uncontrolled B cell proliferation (Sayos et al. 1998; Nichols et al. 1998), encodes a 15-kDa protein, SAP (also known as DSHP), consisting of a single SH2 domain highly homologous to the SH2 domain of SHIP (Nichols et al. 1998). Recent work suggests that CD150 can bind SHIP, SHP-2, or WT SAP, and the presence of WT SAP, which is upregulated in B cells by CD40 crosslinking and downregulated by BCR ligation, facilitates binding of SHIP to CD150. In its absence, SHP-2 binds CD150 (Shlapatska et al. 2001). Thus, perhaps in the presence of mutant SAP, the mitogenic SHP-2, rather than the inhibitory SHIP, binds to CD150 and this may account for the uncontrolled B cell proliferation.

As far as biological ramifications of SHIP are concerned, we have found that SHIP$^{-/-}$ BMMCs are far more prone to degranulation in response to IgE+Ag and, unlike WT BMMCs, degranulate vigorously in response to SF alone (Huber et al. 1998) or to IgE alone (Huber et al. 1998a). We also found that the influx of extracellular calcium is substantially higher in SHIP$^{-/-}$ BMMCs exposed to either IgE alone, IgE+Ag (Huber et al. 1998a) or SF (Huber et al. 1998). In addition, IgE alone, IgE+Ag or SF increases PI-3,4,5-P_3 levels far higher and PI-3,4-P_2 levels significantly less in SHIP$^{-/-}$ than in SHIP$^{+/+}$ BMMCs (Huber et al. 1998; Huber et al. 1999; Scheid et al. 2002). This demonstrates that SHIP and not SHIP2 is the primary enzyme responsible for hydrolyzing IgE- and SF-induced PI-3,4,5-P_3 in normal BMMCs and that a major source of PI-3,4-P_2 in these cells is from PI-3,4,5-P_3. Importantly, we observe no detectable difference in the release of intracellular calcium in SF- or IgE-stimulated SHIP$^{+/+}$ and $^{-/-}$ BMMCs (i.e., in the presence of EGTA), and thus hypothesize, like Bolland et al. (1998), that the elevated PI-3,4,5-P_3 present in SHIP$^{-/-}$ cells attracts and activates a PH-containing intermediate at a step between the draining of intracellular calcium stores and extracellular calcium entry. Thus, SHIP appears to function as a "gatekeeper" in normal BMMCs by keeping PI3K-generated PI-

3,4,5-P_3 levels in check and this restricts extracellular calcium entry and subsequent degranulation. We also propose that SHIP regulates, via PI-3,4,5-P_3 levels, the activation of PDK1 and downstream PKC isoforms that play a role in the cytoskeletal changes important to the degranulation process since many PKC isoforms are substantially elevated at the plasma membrane following SF or IgE-stimulation of SHIP$^{-/-}$ BMMCs (Huber et al. 2000; Kalesnikoff et al. 2002a; Leitges et al. 2002). Related to this, Huber's group has shown recently that PKCδ binds to Shc/SHIP complexes, via the SH2 domain of Shc, and negatively regulates IgE+Ag-induced degranulation, perhaps by facilitating the translocation of SHIP to the plasma membrane (Leitges et al. 2002). Taken together, these results reveal a vital role for SHIP in both setting the threshold for and limiting degranulation of mast cells.

We have also found that IgE- or IgE+Ag-induced inflammatory cytokine production is markedly elevated in SHIP$^{-/-}$ BMMCs and that this is, at least in part, via PI-3,4,5-P_3-mediated activation of NFκB (Kalesnikoff et al. 2002). Related to this, Tridandapani et al. (2002) have found, during phagocytic FcγR activation in human myeloid cells, that SHIP downregulates NFκB-induced gene transcription. These studies indicate that SHIP is a potent inhibitor of the NFκB pathway. In addition, we have strong evidence that IgE- or SF-induced adherence to fibronectin is more rapid and occurs to a greater extent with SHIP$^{-/-}$ than SHIP$^{+/+}$ BMMCs (V. Lam et al., manuscript submitted). These results suggest that SHIP negatively regulates not only degranulation, but cytokine production and adhesion of mast cells as well.

While most SHIP-induced effects are likely mediated by its ability to break down PI-3,4,5-P_3 to PI-3,4-P_2, there is growing evidence that SHIP may also hydrolyze IP_4 in vivo, at least in some cell types (Valderrama-Carvajal et al. 2002) and this could affect the levels of the higher inositol polyphosphates. This in turn may affect protein synthesis levels since it has recently been reported that IP_6 plays an essential role in transporting mRNA out of the nucleus for translation on ribosomes (Feng et al. 2001). Interestingly, there is also growing evidence that PI-3,4-P_2 may act as a second messenger in its own right in some cells by attracting PH-containing proteins, such as the TAPPs (Marshall et al. 2002), that are specifically attracted to this phospholipid (Rameh and Cantley 1999; Jones et al. 1999; Scheid et al. 2002). It is possible therefore that a PTEN knockout phenotype (which leads to an elevation of both PI-3,4,5-P_3 and PI-3,4-P_2) may be qualitatively different from a SHIP or SHIP2 knockout (which elevates PI-3,4,5-P_3 but reduces PI-3,4-P_2). In addition, it is likely that SHIP functions as an adaptor under some circumstances. For example, there is evidence that SHIP (and SHIP2) competes with Grb2 for Shc and thereby reduces Ras activation in some cells (Coggeshall 1998; Ishihara et al. 1999). In addition, there is evidence that during FcγRIIB-mediated inhibition of B cell activation, SHIP also reduces Ras activity by recruiting the RasGAP-binding protein, p62^{Dok} (Tamir et al. 2000). Lastly, SHIP may also play an adaptor role in activating SHP-2 (Koncz et al. 2001).

The role of SHIP downstream of stimulatory cytokines

Although we will be focusing specifically on the role that SHIP plays downstream of stimulatory cytokine receptors in this section, many studies have explored SHIP's role downstream of growth factor receptors (e.g., those for SF, RANK, and M-CSF) and stimulatory immunoreceptors (e.g., FcεRI, BCR, FcγRI, and III) and the reader is directed to several articles and reviews regarding these other extracellular stimuli (Huber et al. 2000; Brauweiler et al. 2000; Liu et al. 2001; Takeshita et al. 2002; Inabe et al. 2002).

Typically, in response to stimulatory cytokines, SHIP becomes tyrosine phosphorylated, associates with Shc and other proteins, and restrains survival and proliferation. For example, we originally showed that this occurred in response to IL-3 in Ba/F3 and DA-ER cells (Liu et al. 1994; Damen et al. 1996; Liu et al. 1997). Subsequently, Drachman and Kaushansky (1997) showed that thrombopoietin (TPO) stimulates SHIP tyrosine phosphorylation via Y^{112} of the TPO receptor (Mpl) in Mpl-expressing Ba/F3 cells. In addition, G-CSF has been shown to stimulate the tyrosine phosphorylation and association of SHIP with Shc in Ba/F3 cells via the distal inhibitory region of a transfected G-CSFR (de Koning et al. 1996; Hunter and Avalos 1998). The generation of a SHIP knockout mouse confirmed SHIP's role as negative regulator of cytokine-stimulated proliferation and survival since bone marrow-derived myeloid progenitors from $SHIP^{-/-}$ mice required far less IL-3 or GM-CSF to proliferate to the same degree as WT cells (Helgason et al. 1998). Consistent with this finding, $SHIP^{-/-}$ mice overproduce mature neutrophils and monocyte/macrophages (Helgason et al. 1998).

However, somewhat ironically, given that we first became interested in SHIP because it was tyrosine phosphorylated in response to erythropoietin (Epo) in Ba/F3 cells expressing exogenous Epo receptors (EpoRs; Liu et al. 1994), $SHIP^{-/-}$ mice do not suffer from polycythemia and are in fact slightly anemic (Helgason et al. 1998; Liu et al. 1998). Consistent with this phenotype, we (Fig. 2A; M.R. Hughes and G. Krystal, unpublished data) and Mason et al. (2002) have found that splenic-derived $SHIP^{-/-}$ erythroid progenitors show a similar or slightly reduced Epo dose response when compared to their $SHIP^{+/+}$ counterparts. To investigate this further, we examined the SHIP protein expression in WT mice as they differentiate down the erythroid lineage and found that late, Epo-responsive, Ter119$^+$ erythroid cells no longer express SHIP (Fig. 2B). Based on this finding, one might expect a similar Epo-dose response for $SHIP^{+/+}$ and $^{-/-}$ Ter119$^+$ cells and the fact that we often observe hyporesponsiveness with the $SHIP^{-/-}$ cells may be due, at least in part, to the presence of negative feedback mechanisms that were upregulated at an earlier stage of erythroid maturation. Consistent with this, it appears that $SHIP^{-/-}$ mice have upregulated a number of negative feedback mechanisms in an attempt to counter the inflammatory effects brought on by their increased PI-3,4,5-P_3 levels (Rauh et al. 2002). The slight anemia may also be due to the fact that myelopoiesis is dramatically elevated in these mice and has taken over the marrow and forced erythropoiesis to occur in the spleen and other extramedullary sites.

Although our SHIP expression studies suggest that SHIP is not expressed in late, Epo-responsive murine erythroid progenitors, this may not be the case in humans (M.R. Hughes and G. Krystal, unpublished data). Thus, studies exploring SHIP's response to Epo stimulation in various human cell lines may still be valid. With this in mind, it has been shown in the human UT-7 cell line that Epo stimulates the tyrosine phosphorylation of the 116-kDa adaptor, Gab1 (and its association with SHIP, PI3K, SHP-2, Shc, and Grb2) and the EpoR-associated IRS-2 (and its association with SHIP and PI3K; Lecoq-Lafon et al. 1999; Verdier et al. 1997). In addition, when Boer et al. (2001) overexpressed WT or catalytically inactive SHIP in the Epo-dependent cell line AS-E2 ,they found, unexpectedly, that the inactive but not the WT form decreased proliferation and resulted in prolonged activation of Erk and PKB. However, when these cells were Epo deprived, they saw increased, caspase 3-dependent apoptosis with WT-SHIP, as one might expect, but

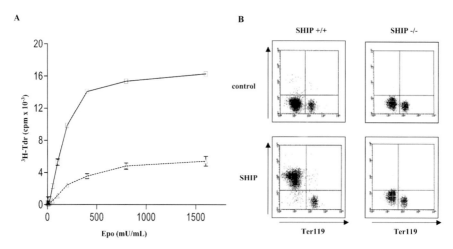

Fig. 2 (**A**) Erythroid progenitors from spleens of phenylhydrazine treated SHIP$^{-/-}$ mice are hypo-responsive to Epo. Cells isolated from the spleens of phenylhydrazine-treated mice were treated with ammonium chloride to remove mature red blood cells and reticulocytes and enriched for erythroid progenitors using a StemSep™ column (anti-Ly-1, anti-B220, anti-Gr-1, and anti-Mac-1). The Epo-induced proliferative response of these cells [□ SHIP$^{+/+}$ (87% Ter119^{+}) and △ SHIP$^{-/-}$ (86% Ter119^{+})] was assessed using a ^{3}H-thymidine proliferation assay and the results shown (mean±SEM of duplicate determinations) are representative of 3 experiments. (**B**) Ter119^{+} erythroid progenitors from WT mice do not express SHIP protein. FACS profile of cells isolated from the spleens of phenylhydrazine-treated SHIP$^{+/+}$ or SHIP$^{-/-}$ mice. Ammonium chloride treated cells were fixed, permeabilized and labeled either with affinity-purified anti-SHIP rabbit polyclonal antibody followed by a PE-coupled anti-rabbit secondary antibody (SHIP, *lower left and right panels*) or with the PE-coupled anti-rabbit secondary antibody alone (control, *upper left and right panels*). The same cells were simultaneously labeled with FITC-coupled anti-Ter119 antibody (all 4 panels)

they found it was independent of its enzymatic activity (Boer et al. 2001). In addition, Siegel et al. (1999) found that SHIP was not expressed in the BCR-ABL-expressing human K562 erythroleukemic cell line and exogenous expression of WT, but not catalytically-inactive, SHIP led to constitutive tyrosine phosphorylation and association with Shc and Grb2 and inhibition of hemin-induced differentiation.

Interestingly, with regard to IL-3, reports by Velazquez et al. (2000) and by Bone and Welham (2000) suggest that Shc, which requires its PTB domain for both its own tyrosine phosphorylation and for its binding to the pY^{577} of the IL-3Rβ chain, brings SHIP to the IL-3R in response to IL-3 stimulation (see model in Fig. 3). Thus, it is very possible that SHIP can be translocated to the plasma membrane by very different mechanisms, depending on the stimulus and cell type. So, for example, it may utilize its own SH2 domain to take it to the FcγRIIB, MAFA, or the FcεRI in mast cells, or use Shc's PTB domain to take it to the IL-3R in mast cells, or its SH2 domain to take it to CD16 in NK cells.

In terms of cytokine-induced downstream signaling events regulated by SHIP, we and others have shown in BMMCs and B cells that SHIP curtails extracellular calcium entry and subsequent plasma membrane localization/activation of both Ca^{++}-dependent and PI-3,4,5-P$_{3}$/PDK1-dependent PKC isoforms (Huber et al. 2000; Chou et al. 1998; Huber et al. 1998; Aman et al. 2000; Kalesnikoff et al. 2002a; Leitges et al. 2002). This is relevant since, as mentioned earlier, Huber's group has shown that PKCδ binds to Shc/SHIP complexes and enhances the negative effects of SHIP following IgE+Ag stimulation of BMMCs (Leitges et al. 2002). Although we typically do not observe an increase in PLCγ or cy-

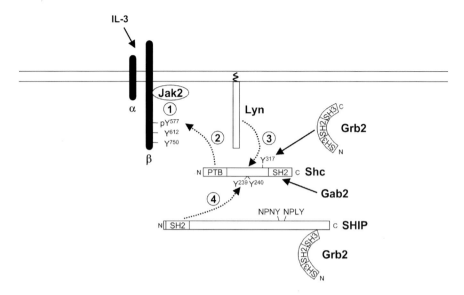

Fig. 3 A Model of IL-3-induced translocation of SHIP to the plasma membrane of mast cells [based in part on studies by Velazquez et al. (2000) and Bone and Welham (2000)]. (**1**) IL-3 stimulates the Jak2-mediated tyrosine phosphorylation of the β_{IL-3} subunit of the IL-3R at Y^{577}. (**2**) This attracts Shc via its PTB domain. (**3**) Shc then gets tyrosine phosphorylated by Lyn or Jak2, primarily at $Y^{239\ and\ 317}$ and (**4**) this attracts SHIP via its SH2 domain. It is likely, as first suggested by Harmer and DeFranco (1999) for B cell activation, that both Shc and Grb2 facillitate the localization of SHIP to the plasma membrane in response to IL-3

tosolic calcium following IL-3 stimulation of single cell suspensions of BMMCs (M. Huber and G. Krystal, unpublished data), the calcium-independent (but DAG-dependent) PKCδ might still play a role here via phospholipase D-generated DAG or via PDK1. Lastly, with regard to IL-3, Liu et al. (1999) showed that IL-3 stimulation of SHIP$^{-/-}$ neutrophils or BMMCs leads to increased and prolonged PI-3,4,5-P$_3$ levels and PKB activation and reduced apoptosis following cytokine removal.

Also worthy of note is that while SHIP is typically thought of as a negative regulator of proliferation/survival and end cell activation, Giallourakis et al. (2000) reported that it may play a positive role in IL-4-induced proliferation. Specifically, they found that overexpressing WT SHIP was hyperproliferative in 32D cells expressing IRS-2 while catalytically inactive SHIP showed reduced proliferation in response to IL-4. However, a more recent paper from the same group (Kashiwada et al. 2001) is somewhat at odds with this finding since they found an ITIM in the IL4Rα chain that binds SHIP and removal of this ITIM also leads to a more rapid proliferation.

The role of SHIP downstream of inhibitory cytokines

Although we will be concentrating on the role of SHIP downstream of negatively acting cytokines in this section, there are a number of excellent articles and reviews dealing with the role of SHIP downstream of other inhibitory receptors (Coggeshall 1998; Scharenberg and Kinet 1998; Bolland et al. 1998).

IL-10 is a key negative regulator of macrophage activation and tumor necrosis factor (TNF)-α production (Moore et al. 2001). It is thought to act by heterodimerizing the IL-10R1 and R2 receptor chains and thus enabling the associated Jak1 and Tyk2 to phosphorylate the IL-10R1 on two tyrosines. This creates docking sites for Stat3, which then becomes phosphorylated and translocates as a dimer into the nucleus to turn on inhibitory genes like p19^{INK4D} (O'Farrell et al. 2000). However, this relatively slow process does not explain the rapid IL-10-induced inhibition of TNF-α production in LPS-stimulated macrophages, and Mui's group has just shown that SHIP is tyrosine phosphorylated in response to IL-10 and plays an essential role in halting TNF-α translation by uncoupling TNF-α mRNA from polysomes and thus stopping its translation (A. Ghanipour et al., manuscript submitted).

Although not strictly members of the cytokine receptor superfamily, it is also worthy of note that death receptors of the TNF/NGF family induce apoptosis at least in part via attraction of SHIP, SHP-1, and SHP-2 to phosphorylated tyrosines within the death domains of these receptors (Daigle et al. 2002). In addition, transforming growth factor (TGF)-β and activin, which are potent inhibitors of hemopoietic cell proliferation and survival, have been shown very recently to exert their effects, at least in part, by upregulating SHIP protein expression (Valderrama-Carvajal et al. 2002). Interestingly, this TGF-β/activin-induced upregulation appears to be limited to SHIP since we have not seen any effects on SHIP2 or PTEN levels (L.M. Sly and G. Krystal, unpublished data) and Valderrana-Carvajal et al. (2002) reported no effect on PTEN expression as well. Interestingly, in this regard, Bruyns et al. (1999) reported that while both SHIP and SHIP2 are expressed in human T lymphocytes, only SHIP2 protein levels are increased after long term stimulation of the T cell receptor. In addition, insulin resistance of diabetic db/db mice appears to be associated with an increase in SHIP2 levels in the skeletal muscle and fat tissue of these mice (Hori et al. 2002). Thus, from these studies it appears that one way for SHIP and SHIP2 to respond to extracellular stimuli is to increase their protein levels. In addition, it was recently reported that high PI-3,4,5-P$_3$ levels lead to PKC-mediated phosphorylation and stabilization of PTEN (Birle et al. 2002). Thus, all three phospholipid phosphatases may feedback inhibit an elevated PI3K pathway, in part, by simply increasing their protein levels.

Future directions

There are still a great many questions to be answered before we can say we have even a rudimentary grasp of how SHIP regulates cytokine-stimulated events. For example, we do not know the role that SHIP's tyrosine phosphorylation plays in its translocation and function(s). In this regard, our studies with a mutant SHIP in which both its NPXYs have been replaced with NPXFs suggest that there are tyrosines in addition to these two that are heavily phosphorylated in response to extracellular stimulation (Damen et al. 2001). Related to this, we do not know the role of the various SHIP-associated proteins in translocating SHIP to the plasma membrane (and cytoskeleton) and whether the presence of SHIP affects the functions of these associated proteins. For example, does the ability of SHIP and sSHIP, but not SHIP2, to bind Grb2 and the ability of SHIP and SHIP2, but not sSHIP, to bind Shc confer nonredundant response capabilities? We also have no idea what the roles of the various SHIP isoforms are and we have not identified with certainty the plasma membrane-associated proteins that attract the various SHIPs. Moreover, it would be very interesting to know the transcription factors that regulate the expression of sSHIP and SHIP, and what regulates the relative levels of the various SHIP isoforms. We are also unclear about the role SHIP plays in regulating IP_4 in vivo in different hemopoietic cells and, in cases where it does, what effect this has on the levels and functions of the higher inositol polyphosphates. Another big question is how extensive a role does SHIP play as an adaptor molecule with regard to the Doks and other molecules. Lastly, in terms of future directions, there is a lot of therapeutic potential in designing small molecular activators and inhibitors of SHIP for the treatment of inflammatory disorders, cancers, and transplantation rejection and we foresee a concerted effort in the near future in this direction.

Acknowledgments. We thank Christine Kelly for typing the manuscript. This work was supported by the NCI-C, with funds from the Terry Fox Foundation, with core support from the BC Cancer Foundation and the BC Cancer Agency.

References

Aman MJ, Walk SF, March ME, Su HP, Carver DJ, Ravichandran KS (2000) Essential role for the C-terminal noncatalytic region of SHIP in FcγRIIB1-mediated inhibitory signaling. Mol Cell Biol 20:3576–3589

Birle D, Bottini N, Williams S, Huynh H, deBelle I, Adamson E, Mustelin T (2002) Negative feedback regulation of the tumor suppressor PTEN by phosphoinositide-induced serine phosphorylation. J Immunol 169:286–291

Boer AK, Drayer AL, Vellenga E (2001) Effects of overexpression of the SH2-containing inositol phosphatase SHIP on proliferation and apoptosis of erythroid AS-E2 cells. Leukemia 15:1750–7175

Bolland S, Pearse RN, Kurosaki T, Ravetch JV (1998) SHIP modulates immune receptor responses by regulating membrane association of Btk. Immunity 8:509–516

Bone H, Welham MJ (2000) Shc associates with the IL-3 receptor β subunit, SHIP and Gab2 following IL-3 stimulation. Contribution of Shc PTB and SH2 domains. Cell Signal 12:183–194

Brauweiler AM, Tamir I, Cambier JC (2000) Bilevel control of B-cell activation by the inositol 5-phosphatase SHIP. Immunol Rev 176:69-74

Brauweiler A, Tamir I, Marschner S, Helgason CD, Cambier JC (2001) Partially distinct molecular mechanisms mediate inhibitory FcγRIIB signaling in resting and activated B cells. J Immunol 167:204–211

Bruyns C, Pesesse X, Moreau C, Blero D, Erneux C (1999) The two SH2-domain-containing inositol 5-phosphatases SHIP1 and SHIP2 are coexpressed in human T lymphocytes. Biol Chem 380:969–974

Cantley LC, Neel BG (1999) New insights into tumor suppression: PTEN suppresses tumor formation by restraining the phosphoinositide 3-kinase/AKT pathway. Proc Natl Acad Sci USA 96:4240–4245

Chou MM, Hou W, Johnson J, Graham LK, Lee MH, Chen CS, Newton AC, Schaffhausen BS, Toker A (1998) Regulation of protein kinase C ζ by PI 3-kinase and PDK-1. Curr Biol 8:1069–1077

Clement S, Krause U, Desmedt F, Tanti JF, Behrends J, Pesesse X, Sasaki T, Penninger J, Doherty M, Malaisse W, et al. (2001) The lipid phosphatase SHIP2 controls insulin sensitivity. Nature 409:92–97

Coggeshall KM (1998) Inhibitory signaling by B cell Fcγ RIIb. Curr Opin Immunol 10:306–312

Daigle I, Yousefi S, Colonna M, Green DR, Simon HU (2002) Death receptors bind SHP-1 and block cytokine-induced antiapoptotic signaling in neutrophils. Nat Med 8:61–67

Damen JE, Liu L, Rosten P, Humphries RK, Jefferson AB, Majerus PW, Krystal G (1996) The 145-kDa protein induced to associate with Shc by multiple cytokines is an inositol tetraphosphate and phosphatidylinositol 3,4,5-trisphosphate 5-phosphatase. Proc Natl Acad Sci USA 93:1689–1693

Damen JE, Liu L, Ware MD, Ermolaeva M, Majerus PW, Krystal G (1998) Multiple forms of the SH2-containing inositol phosphatase, SHIP, are generated by C-terminal truncation. Blood 92:1199–1205

Damen JE, Ware MD, Kalesnikoff J, Hughes MR, Krystal G (2001) SHIP's C-terminus is essential for its hydrolysis of PIP_3 and inhibition of mast cell degranulation. Blood 97:1343–1351

de Koning JP, Schelen AM, Dong F, van Buitenen C, Burgering BM, Bos JL, Lowenberg B, Touw IP (1996) Specific involvement of tyrosine 764 of human granulocyte colony-stimulating factor receptor in signal transduction mediated by p145/Shc/GRB2 or p90/GRB2 complexes. Blood 87:132–140

Drachman JG, Kaushansky K (1997) Dissecting the thrombopoietin receptor: functional elements of the Mpl cytoplasmic domain. Proc Natl Acad Sci USA 94:2350–2355

Dunant NM, Wisniewski D, Strife A, Clarkson B, Resh MD (2000) The phosphatidylinositol polyphosphate 5-phosphatase SHIP1 associates with the dok1 phosphoprotein in bcr-Abl transformed cells. Cell Signal 12:317–326

Dyson JM, O'Malley CJ, Becanovic J, Munday AD, Berndt MC, Coghill ID, Nandurkar HH, Ooms LM, Mitchell CA (2001) The SH2-containing inositol polyphosphate 5-phosphatase, SHIP-2, binds filamin and regulates submembraneous actin. J Cell Biol 155:1065–1079

Feng Y, Wente SR, Majerus PW (2001) Overexpression of the inositol phosphatase SopB in human 293 cells stimulates cellular chloride influx and inhibits nuclear mRNA export. Proc Natl Acad Sci USA 98:875–879

Fox JA, Ung K, Tanlimco SG, Jirik FR (2002) Disruption of a single Pten allele augments the chemotactic response of B lymphocytes to stromal cell-derived factor-1. J Immunol 169:49–54

Galandrini R, Tassi I, Morrone S, Lanfrancone L, Pelicci P, Piccoli M, Frati L, Santoni A (2001) The adaptor protein shc is involved in the negative regulation of NK cell-mediated cytotoxicity. Eur J Immunol 31:2016–2025

Geier SJ, Algate PA, Carlberg K, Flowers D, Friedman C, Trask B, Rohrschneider LR (1997) The human SHIP gene is differentially expressed in cell lineages of the bone marrow and blood. Blood 89:1876–1885

Giallourakis C, Kashiwada M, Pan PY, Danial N, Jiang H, Cambier J, Coggeshall KM, Rothman P (2000) Positive regulation of interleukin-4-mediated proliferation by the SH2-containing inositol-5'-phosphatase. J Biol Chem 275:29275–29282

Giuriato S, Payrastre B, Drayer AL, Plantavid M, Woscholski R, Parker P, Erneux C, Chap H (1997) Tyrosine phosphorylation and relocation of SHIP are integrin-mediated in thrombin-stimulated human blood platelets. J Biol Chem 272:26857–26863

Harmer SL, DeFranco AL (1999) The src homology domain 2-containing inositol phosphatase SHIP forms a ternary complex with Shc and Grb2 in antigen receptor-stimulated B lymphocytes. J Biol Chem 274:12183–12191

Helgason CD, Damen JE, Rosten P, Grewal R, Sorensen P, Chappel SM, Borowski A, Jirik F, Krystal G, Humphries RK (1998) Targeted disruption of SHIP leads to hemopoietic perturbations, lung pathology, and a shortened life span. Genes Dev 12:1610–1620

Hocking LJ, Herbert CA, Nicholls RK, Williams F, Bennett ST, Cundy T, Nicholson GC, Wuyts W, Van Hul W, Ralston SH (2001) Genome-wide search in familial Paget disease of bone shows evidence of genetic heterogeneity with candidate loci on chromosomes 2q36, 10p13, and 5q35. Am J Hum Genet 69:1055–1061

Hori H, Sasaoka T, Ishihara H, Wada T, Murakami S, Ishiki M, Kobayashi M (2002) Association of SH2-containing inositol phosphatase 2 with the insulin resistance of diabetic db/db mice. Diabetes 51:2387–2394

Horn S, Meyer J, Heukeshoven J, Fehse B, Schulze C, Li S, Frey J, Poll S, Stocking C, Jucker M (2001) The inositol 5-phosphatase SHIP is expressed as 145- and 135-kDa proteins in blood and bone marrow cells in vivo, whereas carboxyl-truncated forms of SHIP are generated by proteolytic cleavage in vitro. Leukemia 15:112–120

Huber M, Helgason CD, Scheid MP, Duronio V, Humphries RK, Krystal G (1998) Targeted disruption of SHIP leads to Steel factor-induced degranulation of mast cells. EMBO J 17:7311–7319

Huber M, Helgason CD, Damen JE, Liu L, Humphries RK, Krystal G (1998a) The src homology 2-containing inositol phosphatase (SHIP) is the gatekeeper of mast cell degranulation. Proc Natl Acad Sci USA 95:11330–11335

Huber M, Helgason CD, Damen JE, Scheid M, Duronio V, Liu L, Ware MD, Humphries RK, Krystal G (1999) The role of SHIP in growth factor-induced signaling. Prog Biophys Mol Biol 71:423–434

Huber M, Helgason CD, Damen JE, Scheid MP, Duronio V, Lam V, Humphries RK, Krystal G (1999a) The role of SHIP in FcεRI-induced signaling. In: Daeron M, Vivier E (eds) Current topics in microbiology and immunology, immunoreceptor tyrosine-based inhibition motifs (vol 244), Springer, pp 29–41

Huber M, Damen JE, Ware M, Hughes M, Helgason CD, Humphries RK, Krystal G (2000) Regulation of mast cell degranulation by SHIP. In: Marone G, Lichtenstein LM, Galli SJ (eds) Mast cells and basophils in physiology, pathology and host defense. Academic Press, San Diego, pp 169–182

Hunter MG, Avalos BR (1998) Phosphatidylinositol 3'-kinase and SH2-containing inositol phosphatase (SHIP) are recruited by distinct positive and negative growth-regulatory domains in the granulocyte colony-stimulating factor receptor. J Immunol 160:4979–4987

Inabe K, Ishiai M, Scharenberg AM, Freshney N, Downward J, Kurosaki T (2002) Vav3 modulates B cell receptor responses by regulating phosphoinositide 3-kinase activation. J Exp Med 195:189–200

Ishihara H, Sasaoka T, Hori H, Wada T, Hirai H, Haruta T, Langlois WJ, Kobayashi M (1999) Molecular cloning of rat SH2-containing inositol phosphatase 2 (SHIP2) and its role in the regulation of insulin signaling. Biochem Biophys Res Commun 260:265–272

Johnson GR, Gonda TJ, Metcalf D, Hariharan IK, Cory S (1989) A lethal myeloproliferative syndrome in mice transplanted with bone marrow cells infected with a retrovirus expressing granulocyte-macrophage colony stimulating factor. EMBO J 8:441–448

Jones SM, Klinghoffer R, Prestwich GD, Toker A, Kazlauskas A (1999) PDGF induces an early and a late wave of PI 3-kinase activity, and only the late wave is required for progression through G1. Curr Biol 9:512–521

Kalesnikoff J, Lam V, Krystal G (2002) SHIP represses mast cell activation and reveals that IgE alone triggers signaling pathways which enhance normal mast cell survival. Mol Immunol 38:1201–1206

Kalesnikoff J, Baur N, Leitges M, Hughes MR, Damen JE, Huber M, Krystal G (2002a) SHIP negatively regulates IgE + antigen-induced IL-6 production in mast cells by inhibiting NF-κ B activity. J Immunol 168:4737–4746

Kashiwada M, Giallourakis CC, Pan PY, Rothman PB (2001) Immunoreceptor tyrosine-based inhibitory motif of the IL-4 receptor associates with SH2-containing phosphatases and regulates IL-4-induced proliferation. J Immunol 167:6382–6387

Kavanaugh WM, Pot DA, Chin SM, Deuter-Reinhard M, Jefferson AB, Norris FA, Masiarz FR, Cousens LS, Majerus PW, Williams LT (1996) Multiple forms of an inositol polyphosphate 5-phosphatase form signaling complexes with Shc and Grb2. Curr Biol 6:438–445

Kim CH, Hangoc G, Cooper S, Helgason CD, Yew S, Humphries RK, Krystal G, Broxmeyer HE (1999) Altered responsiveness to chemokines due to targeted disruption of SHIP. J Clin Invest 104:1751–1759

Kimura T, Sakamoto H, Appella E, Siraganian RP (1997) The negative signaling molecule SH2 domain-containing inositol-polyphosphate 5-phosphatase (SHIP) binds to the tyrosine-phosphorylated β subunit of the high affinity IgE receptor. J Biol Chem 272:13991–13996

Koncz G, Toth GK, Bokonyi G, Keri G, Pecht I, Medgyesi D, Gergely J, Sarmay G (2001) Coclustering of Fcγ and B cell receptors induces dephosphorylation of the Grb2-associated binder 1 docking protein. Eur J Biochem 268:3898–3906

Krystal G (2000) Lipid phosphatases in the immune system. Semin Immunol 12:397–403

Lamkin TD, Walk SF, Liu L, Damen JE, Krystal G, Ravichandran KS (1997) Shc interaction with Src homology 2 domain containing inositol phosphatase (SHIP) in vivo requires the Shc-phosphotyrosine binding domain and two specific phosphotyrosines on SHIP. J Biol Chem 272:10396–103401

Lecoq-Lafon C, Verdier F, Fichelson S, Chretien S, Gisselbrecht S, Lacombe C, Mayeux P (1999) Erythropoietin induces the tyrosine phosphorylation of GAB1 and its association with SHC, SHP2, SHIP, and phosphatidylinositol 3-kinase. Blood 93:2578–2585

Leitges M, Gimborn K, Elis W, Kalesnikoff J, Hughes MR, Krystal G, Huber M (2002) Protein kinase C-δ is a negative regulator of antigen-induced mast cell degranulation. Mol Cell Biol 22:3970–3980

Lemay S, Davidson D, Latour S, Veillette A (2000) Dok-3, a novel adapter molecule involved in the negative regulation of immunoreceptor signaling. Mol Cell Biol 20:2743–2754

Lioubin MN, Algate PA, Tsai S, Carlberg K, Aebersold A, Rohrschneider LR (1996) p150Ship, a signal transduction molecule with inositol polyphosphate-5-phosphatase activity. Genes Dev 10:1084–1095

Liu L, Damen JE, Cutler RL, Krystal G (1994) Multiple cytokines stimulate the binding of a common 145-kilodalton protein to Shc at the Grb2 recognition site of Shc. Mol Cell Biol 14 6926–6935

Liu L, Damen JE, Hughes MR, Babic I, Jirik FR, Krystal G (1997) The Src homology 2 (SH2) domain of SH2-containing inositol phosphatase (SHIP) is essential for tyrosine phosphorylation of SHIP, its association with Shc, and its induction of apoptosis. J Biol Chem 272:8983–8988

Liu L, Damen JE, Ware MD, Krystal G (1997a) Interleukin-3 induces the association of the inositol 5-phosphatase SHIP with SHP2. J Biol Chem 272:10998–11001

Liu Q, Shalaby F, Jones J, Bouchard D, Dumont DJ (1998) The SH2-containing inositol polyphosphate 5-phosphatase, ship, is expressed during hematopoiesis and spermatogenesis. Blood 91:2753–2759

Liu Q, Sasaki T, Kozieradzki I, Wakeham A, Itie A, Dumont DJ, Penninger JM (1999) SHIP is a negative regulator of growth factor receptor-mediated PKB/Akt activation and myeloid cell survival. Genes Dev 137:786–791

Liu Y, Jenkins B, Shin JL, Rohrschneider LR (2001) Scaffolding protein Gab2 mediates differentiation signaling downstream of Fms receptor tyrosine kinase. Mol Cell Biol 21:3047–3056

Lucas DM, Rohrschneider LR (1999) A novel spliced form of SH2-containing inositol phosphatase is expressed during myeloid development. Blood 93:1922–1933

Maehama T, Dixon JE (1998) The tumor suppressor, PTEN/MMAC1, dephosphorylates the lipid second messenger, phosphatidylinositol 3,4,5-trisphosphate. J Biol Chem 27322:13375–13378

Marchetto S, Fournier E, Beslu N, Aurran-Schleinitz T, Dubreuil P, Borg JP, Birnbaum D, Rosnet O (1999) SHC and SHIP phosphorylation and interaction in response to activation of the FLT3 receptor. Leukemia 13:1374–1382

Marshall AJ, Krahn AK, Ma K, Duronio V, Hou S (2002) TAPP1 and TAPP2 are targets of phosphatidylinositol 3-kinase signaling in B cells: sustained plasma membrane recruitment triggered by the B-cell antigen receptor. Mol Cell Biol 22:5479–5491

Mason JM, Halupa A, Hyam D, Iscove NN, Dumont DJ, Barber DL (2002) Ship-1 regulates the proliferation and mobilization of the erythroid lineage (abstract). Blood 100:519a

Mikhalap SV, Shlapatska LM, Berdova AG, Law CL, Clark EA, Sidorenko SP (1999) CDw150 associates with src-homology 2-containing inositol phosphatase and modulates CD95-mediated apoptosis. J Immunol 162:5719–5727

Moore KW, de Waal Malefyt R, Coffman RL, O'Garra A (2001) Interleukin-10 and the interleukin-10 receptor. Annu Rev Immunol 19:683–765

Muraille E, Pesesse X, Kuntz C, Erneux C (1999) Distribution of the src-homology-2-domain-containing inositol 5-phosphatase SHIP-2 in both nonhaemopoietic and haemopoietic cells and possible involvement of SHIP-2 in negative signaling of B-cells. Biochem J 342:697–705

Nichols KE, Harkin DP, Levitz S, Krainer M, Kolquist KA, Genovese C, Bernard A, Ferguson M, Zuo L, Snyder E, Buckler AJ, Wise C, Ashley J, Lovett M, Valentine MB, Look AT, Gerald W, Housman DE, Haber DA (1998) Inactivating mutations in an SH2 domain-encoding gene in X-linked lymphoproliferative syndrome. Proc Natl Acad Sci USA 95:13765–13770

O'Farrell AM, Parry DA, Zindy F, Roussel MF, Lees E, Moore KW, Mui AL (2000) Stat3-dependent induction of p19INK4D by IL-10 contributes to inhibition of macrophage proliferation. J Immunol 164:4607–4615

Ono M, Bolland S, Tempst P, Ravetch JV (1996) Role of the inositol phosphatase SHIP in negative regulation of the immune system by the receptor Fc-γ RIIB. Nature 383:263–266

Ono M, Okada H, Bolland S, Yanagi S, Kurosaki T, Ravetch JV (1997) Deletion of SHIP or SHP-1 reveals two distinct pathways for inhibitory signaling. Cell 90:293–301

Osborne MA, Zenner G, Lubinus M, Zhang X, Songyang Z, Cantley LC, Majerus P, Burn P, Kochan JP (1996) The inositol 5'-phosphatase SHIP binds to immunoreceptor signaling motifs and responds to high affinity IgE receptor aggregation. J Biol Chem 271:29271–29278

Ott VL, Tamir I, Niki M, Pandolfi PP, Cambier JC (2002) Downstream of kinase, p62^{dok}, is a mediator of FcγIIB inhibition of FcεRI signaling. J Immunol 168:4430–4439

Pesesse X, Deleu S, De Smedt F, Drayer L, Erneux C (1997) Identification of a second SH2-domain-containing protein closely related to the phosphatidylinositol polyphosphate 5-phosphatase SHIP. Biochem Biophys Res Commun 239:697–700

Pesesse X, Moreau C, Drayer AL, Woscholski R, Parker P, Erneux C (1998) The SH2 domain containing inositol 5-phosphatase SHIP2 displays phosphatidylinositol 3,4,5-trisphosphate and inositol 1,3,4,5-tetrakisphosphate 5-phosphatase activity. FEBS Lett 437:301–303

Phee H, Jacob A, Coggeshall KM (2000) Enzymatic activity of the Src homology 2 domain-containing inositol phosphatase is regulated by a plasma membrane location. J Biol Chem 275:19090–19097

Rameh LE, Cantley LC (1999) The role of phosphoinositide 3-kinase lipid products in cell function. J Biol Chem 27413:8347–8350

Rauh MJ, Pereira C, Palmer J, Damen J, Mui ALF, Krystal G (2002) SHIP-deficiency leads to anti-inflammatory macrophage programming and endotoxin tolerance (abstract). Blood 100:147a

Sattler M, Salgia R, Shrikhande G, Verma S, Choi JL, Rohrschneider LR, Griffin JD (1997) The phosphatidylinositol polyphosphate 5-phosphatase SHIP and the protein tyrosine phosphatase SHP-2 form a complex in hematopoietic cells which can be regulated by BCR/ABL and growth factors. Oncogene 15:2379–2384

Sattler M, Verma S, Pride YB, Salgia R, Rohrschneider LR, Griffin JD (2001) SHIP1, an SH2 domain containing polyinositol-5-phosphatase, regulates migration through two critical tyrosine residues and forms a novel signaling complex with DOK1 and CRKL. J Biol Chem 276:2451–2458

Sayos J, Wu C, Morra M, Wang N, Zhang X, Allen D, van Schaik S, Notarangelo L, Geha R, Roncarolo MG, Oettgen H, de Vries JE, Aversa G, Terhorst C (1998) The X-linked lymphoproliferative-disease gene product SAP regulates signals induced through the coreceptor SLAM. Nature 395:462–469

Scharenberg AM, Kinet JP (1998) PtdIns-3,4,5-P3: a regulatory nexus between tyrosine kinases and sustained calcium signals. Cell 94:5–8

Scheid MP, Huber M, Damen JE, Hughes M, Kang V, Neilsen P, Prestwich GD, Krystal G, Duronio V (2002) Phosphatidylinositol(3,4,5)P$_3$ is essential but not sufficient for protein kinase B (PKB) activation: Phosphatidylinositol(3,4)P$_2$ is required for PKB phosphorylation at Ser473. Studies using cells from SH2-containing inositol-5-phosphatase knockout mice. J Biol Chem 277:9027–9035

Shlapatska LM, Mikhalap SV, Berdova AG, Zelensky OM, Yun TJ, Nichols KE, Clark EA, Sidorenko SP (2001) CD150 association with either the SH2-containing inositol phosphatase or the SH2-containing protein tyrosine phosphatase is regulated by the adaptor protein SH2D1A. J Immunol 166:5480–5487

Siegel J, Li Y, Whyte P (1999) SHIP-mediated inhibition of K562 erythroid differentiation requires an intact catalytic domain and Shc binding site. Oncogene 18:7135–7148

Stambolic V, Suzuki A, de la Pompa JL, Brothers GM, Mirtsos C, Sasaki T, Ruland J, Penninger JM, Siderovski DP, Mak TW (1998) Negative regulation of PKB/Akt-dependent cell survival by the tumor suppressor PTEN. Cell 95:29–39

Takeshita S, Namba N, Zhao JJ, Jiang Y, Genant HK, Silva MJ, Brodt MD, Helgason CD, Kalesnikoff J, Rauh MJ, Humphries RK, Krystal G, Teitelbaum SL, Ross FP (2002) SHIP-deficient mice are severely osteoporotic due to increased numbers of hyperresorptive osteoclasts. Nat Med 8:943–949

Tamir I, Stolpa JC, Helgason CD, Nakamura K, Bruhns P, Daeron M, Cambier JC (2000) The RasGAP-binding protein p62dok is a mediator of inhibitory FcγRIIB signals in B cells. Immunity 12:347–358

Tridandapani S, Kelley T, Pradhan M, Cooney D, Justement LB, Coggeshall KM (1997) Recruitment and phosphorylation of SH2-containing inositol phosphatase and Shc to the B-cell Fcγ immunoreceptor tyrosine-based inhibition motif peptide motif. Mol Cell Biol 17:4305–4311

Tridandapani S, Pradhan M, LaDine JR, Garber S, Anderson CL, Coggeshall KM (1999) Protein interactions of Src homology 2 (SH2) domain-containing inositol phosphatase (SHIP): association with Shc displaces SHIP from FcγRIIb in B cells. J Immunol 162:1408–1414

Tridandapani S, Wang Y, Marsh CB, Anderson CL (2002) Src homology 2 domain-containing inositol polyphosphate phosphatase regulates NF-κB-mediated gene transcription by phagocytic FcγRs in human myeloid cells. J Immunol 169:4370–4378

Tu Z, Ninos JM, Ma Z, Wang JW, Lemos MP, Desponts C, Ghansah T, Howson JM, Kerr WG (2001) Embryonic and hematopoietic stem cells express a novel SH2-containing inositol 5'-phosphatase isoform that partners with the Grb2 adapter protein. Blood 98:2028–2038

Valderrama-Carvajal H, Cocolakis E, Lacerte A, Lee EH, Krystal G, Ali S, Lebrun JJ (2002) Activin/TGF-β induce apoptosis through Smad-dependent expression of the lipid phosphatase SHIP. Nat Cell Biol 4:963–969

van Dijk TB, van Den Akker E, Amelsvoort MP, Mano H, Lowenberg B, von Lindern M (2000) Stem cell factor induces phosphatidylinositol 3'-kinase-dependent Lyn/Tec/Dok-1 complex formation in hematopoietic cells. Blood 96:3406–3413

Velazquez L, Gish GD, van Der Geer P, Taylor L, Shulman J, Pawson T (2000) The shc adaptor protein forms interdependent phosphotyrosine-mediated protein complexes in mast cells stimulated with interleukin 3. Blood 96:132–138

Vely F, Olivero S, Olcese L, Moretta A, Damen JE, Liu L, Krystal G, Cambier JC, Daeron M, Vivier E (1997) Differential association of phosphatases with hematopoietic coreceptors bearing immunoreceptor tyrosine-based inhibition motifs. Eur J Immunol 27:1994–2000

Verdier F, Chretien S, Billat C, Gisselbrecht S, Lacombe C, Mayeux P (1997) Erythropoietin induces the tyrosine phosphorylation of insulin receptor substrate-2. An alternate pathway for erythropoietin-induced phosphatidylinositol 3-kinase activation. J Biol Chem 272:26173–26178

Ware MD, Rosten P, Damen JE, Liu L, Humphries RK, Krystal G (1996) Cloning and characterization of the human 145 kDa SHC-associated inositol 5-phosphatase, SHIP. Blood 88:2833–2840

Wisniewski D, Strife A, Swendeman S, Erdjument-Bromage H, Geromanos S, Kavanaugh WM, Tempst P, Clarkson B (1999) A novel SH2-containing phosphatidylinositol 3,4,5-trisphosphate 5-phosphatase (SHIP2) is constitutively tyrosine phosphorylated and associated with src homologous and collagen gene (SHC) in chronic myelogenous leukemia progenitor cells. Blood 93:2707–2720

Wolf I, Lucas DM, Algate PA, Rohrschneider LR (2000) Cloning of the genomic locus of mouse SH2 containing inositol 5-phosphatase (SHIP) and a novel 110-kDa splice isoform, SHIPδ. Genomics 69:104–112

Xu R, Abramson J, Fridkin M, Pecht I (2001) SH2 domain-containing inositol polyphosphate 5'-phosphatase is the main mediator of the inhibitory action of the mast cell function-associated antigen. J Immunol 167:6394–6402

Instructions for authors

1 Legal requirements

The author(s) guarantee(s) that the manuscript will not be published elsewhere in any language without the consent of the copyright holders, that the rights of third parties will not be violated, and that the publisher will not be held legally responsible should there be any claims for compensation.

Authors wishing to include figures or text passages that have already been published elsewhere are required to obtain permission from the copyright holder(s) and to include evidence that such permission has been granted when submitting their papers. Any material received without such evidence will be assumed to originate from the authors.

Manuscripts must be accompanied by the "Copyright Transfer Statement".

Please include at the end of the acknowledgements a declaration that the experiments comply with the current laws of the country in which they were performed.

2 Editorial procedure

Manuscripts should be submitted in English, together with one set of illustrations and a complete pdf file, to the editor in charge.

The author is responsible for the accuracy of the references.

3 Manuscript preparation

To help you prepare your manuscript, Springer offers a template that can be used with Winword 7 (Windows 95), Winword 6 and Word for Macintosh.

For details see point 4.

All manuscripts are subject to copy editing.

- **Title page**
  - The name(s) of the author(s)
  - A concise and informative title
  - The affiliation(s) and address(es) of the author(s)
  - The e-mail address, telephone and fax numbers of the communicating author

- **Abstract.** Each paper must be preceded by an abstract presenting the most important results and conclusions.

- **Abbreviations** should be defined at first mention in the abstract and again in the main body of the text and used consistently thereafter

A list of **symbols** should follow the abstract if such a list is needed. Symbols must be written clearly. The international system of units (SI units) should be used. The numbering of chapters should be in decimal form.

Footnotes on the title page are not given reference symbols. Footnotes to the text are numbered consecutively; those to tables should be indicated by superscript lower-case letters (or asterisks for significance values and other statistical data).

Acknowledgements. These should be as brief as possible. Any grant that requires acknowledgement should be mentioned. The names of funding organizations should be written in full.

Funding. Authors are expected to disclose any commercial or other associations that might pose a conflict of interest in connection with submitted material. All funding sources supporting the work and institutional or corporate affiliations of the authors should be acknowledged.

- **References**

The list of References should only include works that are cited in the text and that have been published or accepted for publication. Personal communications should only be mentioned in the text.

In the text, references should be cited by author and year (e.g. Hammer 1994; Hammer and Sjöqvist 1995; Hammer et al. 1993) and listed in alphabetical order in the reference list.

Examples:

Monographs:
Snider T, Grand L (1982) Air pollution by nitrogen oxides. Elsevier, Amsterdam

Anthologies and proceedings:
Noller C, Smith VR (1997) Ultraviolet selection pressure on earliest organisms. In: Kingston H, Fulling CP (eds) Natual environment background analysis. Oxford University Press, Oxford, pp 211–219

Journals:
Meltzoff AN, Moore MK (1977) Imitation of facial and manual gestures by human neonates. Science 198:75–78

If available the Digital Object Identifier (DOI) of the cited literature should be added at the end of the reference in question.

- **Illustrations and Tables**

All figures (photographs, graphs or diagrams) and tables should be cited in the text, and each numbered consecutively throughout. Figure parts should be identified by lower-case roman letters. The placement of figures and tables should be indicated in the left margin. For submission of figures in electronic form see below

Line drawings. Please submit good-quality prints. The inscriptions should be clearly legible.

Half-tone illustrations (black and white and color). Please submit well-contrasted photographic prints with the top indicated on the back. Magnification should be indicated by scale bars.

Figure legends must be brief, self-sufficient explanations of the illustrations. The legends should be placed at the end of the text.

Tables should have a title and a legend explaining any abbreviation used in that table. Footnotes to tables should be indicated by superscript lower-case letters (or asterisks) for significance values and other statistical data.

4 Electronic submission of final version

Please send only the final version of the article, as accepted by the editors.

Preparing your manuscript

The template is available:

→ via ftp:
  Address: ftp.springer.de/
  User ID: ftp
  Password: your own e-mail address
  – Directory: /pub/Word/journals
  – File names: either sv-journ.zip or sv-journ.doc and sv-journ.dot

→ via browser
  – http://www.springer.de/author/index.html

The zip file should be sent uuencoded.

Layout guidelines
1. Use a normal, plain font (e.g., Times Roman) for text.
   Other style options:
   – for textual emphasis use italic types.
   – for special purposes, such as for mathematical vectors, use boldface type.
2. Use the automatic page numbering function to number the pages.
3. Do not use field functions.
4. For indents use tab stops or other commands, not the space bar.
5. Use the table functions of your word processing program, not spreadsheets, to make tables.
6. Use the equation editor of your word processing program or MathType for equations.
7. Place any figure legends or tables at the end of the manuscript.
8. Submit all figures as separate files and do not integrate them within the text.

Data formats
Save your file in two different formats:

1. RTF (Rich Text Format) or Word compatible Word 95/97
2. pdf (a single pdf file including text, tables and figures)

Illustrations

The preferred figure formats are EPS for vector graphics exported from a drawing program and TIFF for halftone illustrations. EPS files must always contain a preview in TIFF of the figure. The file name (one file for each figure) should include the figure number. Figure legends should be included in the text and not in the figure file.

Scan resolution: Scanned line drawings should be digitized with a minimum resolution of 800 dpi relative to the final figure size. For digital halftones, 300 dpi is usually sufficient.

Color illustrations: Store color illustrations as RGB (8 bits per channel) in TIFF format.

General information on data delivery

Please send us a zip file (text and illustrations in separate files) either:

→ Via ftp.springer.de
  (to our ftp.server; log-in "anonymous"; password: your e-mail address; further information in the readme file on the server)

→ By e-mail
  (only suitable for small volumes of data)
→ or on any of the following media:
  – On a diskette [you may use .tar, .zip, .gzip (.gz), .sit, and compress (.Z)]
  – On a ZIP cartridge
  – On a CD-ROM

Please always supply the following information with your data: journal title, operating system, word processing program, drawing program, image processing program, compression program.

The file name should be memorable (e.g., author name), have no more than 8 characters, and include no accents or special symbols. Use only the extensions that the program assigns automatically.

5 Proofreading

Authors should make their proof corrections on a printout of the pdf file supplied, checking that the text is complete and that all figures and tables are included. After online publication, further changes can only be made in the form of an Erratum, which will be hyperlinked to the article. The author is entitled to formal corrections only. Substantial changes in content, e.g. new results, corrected values, title and authorship are not allowed without the approval of the editor in charge. In such a case please contact the Editor in charge before returning the proofs to the publisher.

6 Offprints, Free copy

You are entitled to receive a pdf file of your article for your own personal use. Orders for offprints can be placed by returning the order form with the corrected proofs. One complimentary copy of the issue in which your article appears is supplied.

Printing: Saladruck Berlin
Binding Lüderitz&Bauer, Berlin